Ninian Glen

Actuarial Science

An elementary manual

Ninian Glen

Actuarial Science
An elementary manual

ISBN/EAN: 9783337035792

Printed in Europe, USA, Canada, Australia, Japan

Cover: Foto ©berggeist007 / pixelio.de

More available books at **www.hansebooks.com**

ACTUARIAL SCIENCE

AN ELEMENTARY MANUAL

BY

NINIAN GLEN, M.A., C.A.,

FELLOW OF THE FACULTY OF ACTUARIES.

GLASGOW:

JOHN SMITH & SON, 19 RENFIELD STREET.

1893.

PREFACE.

THIS little handbook is the outcome of a course of lectures delivered by the author during the winter of 1891-92, under the auspices of the Institute of Accountants and Actuaries in Glasgow. It is intended as an elementary treatise on Actuarial Science, and is published in the hope that it may, in part at least, supply a want which has frequently been expressed. The existing Text Books are all that could be desired for the professional Actuary, but they are much too advanced for those who make the subject only a side branch of study.

The author puts forward no claim to originality in connection with this treatise. The ground it covers has frequently been traversed before. If he has succeeded in stating clearly the elementary principles of the science, and in smoothing to some extent the path of the student, his purpose has been amply fulfilled.

CONTENTS.

PART I.—INTEREST AND ANNUITIES-CERTAIN.

PART II.—LIFE CONTINGENCIES.

CHAPTER I.

CHAPTER II.

CHAPTER III.

CHAPTER IV.

CHAPTER V.

CHAPTER VI.

APPENDIX.

Interest.

i = effective rate of interest on £1 for one year.

j = nominal rate of interest on £1 for one year.

n = number of years.

v = present value of 1 due in 1 year = $(1+i)^{-1}$.

v^n = present value of 1 due in n years = $(1+i)^{-n}$.

d = discount on 1 for one year = $1 - v = vi$.

Annuities-Certain

$s_{\overline{n}|}$ = amount of an annuity-certain of 1 for n years.

$a_{\overline{n}|}$ = present value of an annuity-certain of 1 for n years.

$\ddot{a}_{\overline{n}|}$ = present value of an annuity-due of 1 for n years.

a_{∞} = present value of a perpetuity of 1.

$t/a_{\overline{n}|}$ = present value of an annuity-certain of 1 for n years deferred t years.

$a_{\overline{n}|}^{m}$ = present value of an annuity-certain of 1 for n years, payable m times a year.

Life Contingencies.

l_x = number living at exact age x.

d_x = number dying between ages x and $x+1$.

L_x = population of age x, = $\dfrac{l_x + l_{x+1}}{2}$.

T_x = total population from age x upwards, = $L_x + L_{x+1} + L_{x+2}$ + etc.

p_x = probability of a person aged x living one year.

$_np_x$ = probability of a person aged x living n years.

q_x = probability of a person aged x dying within one year.

$'_nq_x$ = probability of a person aged x dying within n years.

e_x = curtate expectation of life of a person aged x.

$\overset{o}{e}_x$ = complete expectation of life of a person aged x.

$$D_x = v^x l_x \qquad\qquad C_x = v^{x+1} d_x$$

$$N_x = D_{x+1} + D_{x+2} + \text{etc.} \qquad M_x = C_x + C_{x+1} + \text{etc.}$$

$$S_x = N_x + N_{x+1} + \text{etc.} \qquad R_x = M_x + M_{x+1} + \text{etc.}$$

Life Benefits.

$_nE_x$ = value of an endowment of 1 payable in n years.

a_x = value of a life annuity of 1 to a person aged x.

$\mathfrak{a}_x$ = value of an annuity-due of 1 to a person aged x.

$_{n/}a_x$ = value of an annuity to a person aged x deferred n years.

$_{/n}a_x$ = value of a temporary annuity for n years to a person aged x.

a_{xy} = value of an annuity during the joint continuance of two lives.

$a_{y/x}$ = value of an annuity to a person aged x after the death of another aged y.

$\overset{\circ}{a}_x$ = value of a complete annuity.

A_x = value of an assurance of 1 payable six months after the death of a person aged x.

$_{n/}A_x$ = value of an assurance deferred n years.

$_{/n}A_x$ = value of a temporary assurance for n years

P_x = value of the net annual premium for an assurance of 1.

Note.—The above symbols are in accordance with the system of notation approved by the Institute of Actuaries, and almost universally adopted.

INTEREST AND ANNUITIES-CERTAIN.

PART I.

INTEREST AND ANNUITIES-CERTAIN.

CHAPTER I.

INTEREST.

ACCUMULATION AT SIMPLE AND COMPOUND INTEREST. PRESENT VALUES AND DISCOUNT.

THERE are two methods of calculating interest.

Simple and Compound Interest. (1) It may be charged upon the original loan or principal only. This is known as *Simple Interest.*

(2) The principal and accrued interest may be accumulated at periodical intervals, and interest charged on the amount. This is *Compound Interest* as generally understood.

Simple Interest, while very common in commercial transactions, is practically never employed in actuarial calculations. It is an illogical process, and leads to unsatisfactory results. Indeed, the true and logical idea of interest is not even Compound Interest as we have above defined it, but rather a momently growth of the principal under the operation of interest. It is in this direction that the more recent developments of Actuarial Science are tending, but the idea can only be worked out with the help of the higher mathematics.

In dealing with Compound Interest, as above defined, much will evidently depend on the periods at which principal and interest are accumulated. This may take place once a year, once every six months, once every three months, or even at more frequent intervals. If, for instance, we are working with a nominal rate of interest of say 5 per cent. per annum, it is obvious that while 5 per cent. will be the actual rate of interest earned, if

A

principal and interest be compounded once a year, the actual or *effective* rate of interest will not be 5 per cent., but something higher, if the accumulation take place half-yearly or quarterly. In point of fact, if the conversion be made quarterly, a nominal annual rate of 5 per cent. is equivalent to an actual or effective rate of 5·095 per cent. The exact relations that exist between the nominal and effective rates of interest will shortly be explained.

We now proceed to illustrate the different problems that arise in connection with interest: and, to obtain general expressions, let—

P = The Principal..

S = The Amount to which the principal will accumulate under the operation of interest.

n = The Number of years.

i = The Interest on £1 for 1 year.

First, and briefly, in the case of *Simple Interest*:—

Accumulation at Simple Interest. The interest on £1 for one year being i, the interest on any sum, P, will be iP, and for n years niP. We have, therefore,

S (the accumulated amount) = P (the principal) + the interest upon it, that is

$$S = P + niP = P(1 + ni),$$

so that, given P, i, and n, S can at once be found.

If S, i, and n be given, to find P we transpose the above equation, thus

$$P = \frac{S}{1 + ni}.$$

Similarly, to find n and i, we have

$$n = \frac{S - P}{iP}, \text{ and } i = \frac{S - P}{nP}.$$

Thus, given any three of the quantities, P, S, n, and i, we can at once find the fourth.

It is not necessary to go further into the question of Simple Interest, as it is of comparatively little importance.

Accumulation at Compound Interest. Let us now consider the same problems in *Compound Interest*—accumulating principal and interest once a year.

The amount to which 1 will accumulate in one year under the operation of interest is manifestly $(1 + i)$, and we start the second year with this as principal, so that at the end of the second year we have $(1+i)(1+i)$ or $(1+i)^2$. We start the third year with $(1+i)^2$ as principal, and at the end of the year it will have accumulated to $(1+i)^2(1+i)$ or $(1+i)^3$. The amount, therefore, to which 1 will accumulate in n years (to put the case generally) is $(1+i)^n$.

Having thus ascertained the amount to which 1 will accumulate, the amount to which any other principal, say P, will accumulate is found by simple multiplication, and calling this amount S, we obtain the general formula applicable in all cases,

$$S = P(1+i)^n.$$

If S, i, and n be given, and it be desired to find P—transposing the above equation

$$P = \frac{S}{(1+i)^n}.$$

Again, if P, S, and i, be given, to find n, we have, as above,

$$S = P(1+i)^n,$$
$$\text{whence, } \log S = \log P + n \log (1+i),$$
$$\text{and, } n \log (1+i) = \log S - \log P,$$
$$\therefore n = \frac{\log S - \log P}{\log (1+i)}.$$

Lastly, given S, P, and n to find i—

$$S = P(1+i)^n,$$
$$\text{whence, } (1+i)^n = \frac{S}{P},$$
$$\therefore (1+i) = \left(\frac{S}{P}\right)^{\frac{1}{n}},$$
$$\text{and, } i = \left(\frac{S}{P}\right)^{\frac{1}{n}} - 1.$$

Nominal and
Effective Rates
of Interest
In arriving at the above formulae, we have proceeded on the assumption that interest is accumulated with the principal only once a year; let us now see in what way these formulae are affected by the more frequent accumulation of principal and interest. This brings us to consider the relation between the *nominal* and *effective* rates of interest, to which reference has already been made.

Let us suppose that j is the nominal annual rate of interest, and that it is added to the principal m times in the year.

By reasoning similar to that already employed, it is evident that the original 1 invested will have accumulated at the end of the first m^{th} part of the year to $\left(1 + \frac{j}{m}\right)$; at the end of the second m^{th} part to $\left(1 + \frac{j}{m}\right)^2$; at the end of the third m^{th} part to $\left(1 + \frac{j}{m}\right)^3$, and at the end of the year to $\left(1 + \frac{j}{m}\right)^m$. Deducting from this the 1 originally invested, we have, as the *interest* realised in the year $\left\{\left(1 + \frac{j}{m}\right)^m - 1\right\}$, which is the effective rate corresponding to the nominal rate j. If, therefore, the effective rate of interest be denoted by i, we have

$$i = \left\{\left(1 + \frac{j}{m}\right)^m - 1\right\}.$$

Conversely, to find the nominal rate of interest in terms of the effective rate—transposing the above equation

$$1 + i = \left(1 + \frac{j}{m}\right)^m,$$

whence, $(1 + i)^{\frac{1}{m}} - 1 = \frac{j}{m}$,

and, $j = m\left\{(1 + i)^{\frac{1}{m}} - 1\right\}.$

The numerical difference between the nominal and effective rates of interest will be seen from the following tables, which are based upon the above formulae. Table I. gives the effective rates corresponding to given nominal rates, and Table II. the nominal rates corresponding to given effective rates.

	TABLE I.			TABLE II.	
NOMINAL RATES, Per Cent.	CORRESPONDING EFFECTIVE RATES when Interest accumulated		EFFECTIVE RATES, Per Cent.	CORRESPONDING NOMINAL RATES when Interest accumulated	
	Half-Yearly.	Quarterly.		Half-Yearly.	Quarterly.
3	3 0225	3·0339	3	2·9778	2·9668
$3\frac{1}{2}$	3·5306	3·5462	$3\frac{1}{2}$	3·4699	3·4550
4	4·0400	4·0604	4	3·9608	3·9414
$4\frac{1}{2}$	4·5506	4·5765	$4\frac{1}{2}$	4·4505	4·4260
5	5·0625	5·0945	5	4·9390	4·9089
6	6·0900	6·1364	6	5 9126	5·8696

Let us now turn to the expressions already found for S, P, i, and n. If the given rate of interest be the effective rate, the formulae of course need no alteration; but if j, the nominal rate of interest be given, and if the interest be converted m times a year, we must alter the formulae as follows :—

$$S = P\left(1 + \frac{j}{m}\right)^{mn},$$

$$P = \frac{S}{\left(1 + \frac{j}{m}\right)^{mn}},$$

$$n = \frac{\log S - \log P}{m \log\left(1 + \frac{j}{m}\right)},$$

$$j = m\left\{\left(\frac{S}{P}\right)^{\frac{1}{mn}} - 1\right\}.$$

There is a pitfall into which beginners are very apt to stumble in putting these formulae into practice, to which it may be well to refer. The rate of interest is usually stated at so much *per cent.*, but it must be remembered that i represents the interest on 1, not on 100. If, therefore, the rate be say four per cent., i is not 4 but ·04. This is a very simple matter, but nevertheless it is one about which the beginner is very apt to go wrong.

We may now summarize the results we have found as follows :—

SIMPLE INTEREST.	COMPOUND INTEREST ACCRUING YEARLY	COMPOUND INTEREST ACCRUING m TIMES A YEAR. j=Nominal Rate of Interest.
$S = P(1 + ni)$.	$S = P(1 + i)^n$.	$S = P\left(1 + \frac{j}{m}\right)^{mn}$.
$P = \dfrac{S}{1 + ni}$.	$P = \dfrac{S}{(1 + i)^n}$.	$P = \dfrac{S}{\left(1 + \frac{j}{m}\right)^{mn}}$.
$n = \dfrac{S - P}{iP}$.	$n = \dfrac{\log S - \log P}{\log (1 + i)}$.	$n = \dfrac{\log S - \log P}{m \log\left(1 + \frac{j}{m}\right)}$.
$i = \dfrac{S - P}{nP}$.	$i = \left(\dfrac{S}{P}\right)^{\frac{1}{n}} - 1$.	$j = m\left\{\left(\dfrac{S}{P}\right)^{\frac{1}{mn}} - 1\right\}$.

Present
Values. In arriving at these formulae we have defined P as the principal, and S as the amount to which it will accumulate under the operation of interest, but the matter must be now looked at from a different standpoint.

If S be the principal to be received on the expiry of n years, P, from this point of view, is evidently its Present Value.

We have found above that $P = \dfrac{S}{(1+i)^n}$; and that this expression correctly represents the present value of the sum S, due at the end of n years, can easily be seen. Since $(1+i)$ is the amount to which 1 accumulates in one year, 1 is manifestly the present value of $(1+i)$ due a year hence, and by simple proportion the present value of 1 due a year hence is $\dfrac{1}{1+i}$. Similarly, $(1+i)^n$ being the amount of 1 in n years, 1 is the present value of $(1+i)^n$ due at the end of n years, and $\dfrac{1}{(1+i)^n}$ the present value of 1 due at the end of n years.

The present value of 1 due at the end of a year is now universally represented by the symbol v, so that we have

$$v = \frac{1}{1+i}, \text{ and } v^n = \frac{1}{(1+i)^n}.$$

Discount. The difference between a sum due at a future date and its present value is called *Discount*.

Using for discount the symbol d, we accordingly have, by definition,

$$d = 1 - v = 1 - \frac{1}{1+i} = \frac{i}{1+i} = iv.$$

In ordinary commercial transactions the practice is simply to deduct as discount, the interest on the principal for the time that will elapse till it becomes payable. In discounting at 5 per cent. an obligation for £100, due at the end of a year, a banker would accordingly deduct £5 and pay over £95. That this method of calculating discount is incorrect is easily seen. Let us suppose (to take an extreme case) that the amount due under the obligation had not been payable for 25 years. Calculated in this way the discount would amount to £125, and we would have the absurd result, that £100 due at the end of 25 years, is worth £25 less than nothing. Discount calculated in this way is known as Trade Discount or Commercial Discount, to distinguish it from True Discount which, as we have seen, is the difference between a sum due and its present value.

The true discount on £100, due at the end of 1 year, at 5 per cent. is $\left(100 \times \dfrac{\cdot05}{1\cdot05}\right)$, and the true discount on £100, due at the end of 25 years, is $\left(100 \times \dfrac{25 \times \cdot05}{1 + 25 \times \cdot05}\right)$, or $\left\{100 \times \left(1 - \dfrac{1}{1\cdot05^{25}}\right)\right\}$, according as it is calculated at simple or compound interest.

Construction of Compound Interest Tables. To save the labour of making isolated calculations at Compound Interest, Tables of Amounts and Present Values are employed when available, and it may sometimes even be advisable to construct such tables when they cannot otherwise be obtained.

In the construction of tables of Compound Interest (and indeed of actuarial tables generally) it is always desirable to have

(1) A direct and satisfactory means of finding the initial value of the function to be tabulated;

(2) A continuous working formula by which each succeeding value can be found from the preceding one; and

(3) A verification formula, by which the accuracy of the previous work can, from time to time, be ascertained by an independent calculation.

The advantages of working in this way are very obvious. If each succeeding value be calculated independently, there are no means of eliminating errors, other than checking and re-checking the work. But, if the second value be obtained from the first, the third from the second, and so on, and if, say at every tenth value, the work be checked by an independent calculation, the accuracy is proved not only of the particular value checked but of all preceding values.

The Table of Amounts is easily formed in this way. The initial value is, of course, $(1 + i)$, and by continuous multiplication by $(1 + i)$ the succeeding values are obtained. For verification, say every tenth value can be calculated by logarithms.

The Table of Present Values is more troublesome to calculate, and can be most simply formed by beginning with the last value—say v^n—and working backwards, multiplying each value by $(1 + i)$ to get the preceding one, since $v^{n-1} = v^n(1 + i)$. For verification in this case also, an independent calculation by logarithms can be made at intervals.

A table may be effectively used for longer terms of years than it actually includes. We can, for example, easily find the values of $(1+i)^{70}$ and v^{70} from a table which goes to 50 years only, as $(1+i)^{70} = (1+i)^{50} \times (1+i)^{20}$, and $v^{70} = v^{50} \times v^{20}$.

A yearly table can likewise frequently be used when the interest is convertible half-yearly or quarterly. Thus the amount of 1 accumulated for 10 years at 5 per cent., convertible half-yearly, is the same as the amount of 1 for 20 years at $2\frac{1}{2}$ per cent. convertible yearly. The same holds good with tables of Present Values.

Practical Examples.

(1) B returns to A £514 5s. for the use of a sum for 7 years which A lent him at 3 per cent. simple interest. What was the sum originally advanced?

Here the formula is $P = \dfrac{S}{1+ni}$, and as $S = 514.25$, $n = 7$, and $i = .03$, we have $P = \dfrac{514.25}{1.21} = £425$—the amount of the original advance.

(2) To what sum will £40 accumulate in 25 years at 5 per cent. compound interest payable yearly?

The formula here is $S = P(1+i)^n$, and as $P = 40$, $i = .05$, and $n = 25$, we have

$$S = 40(1.05)^{25}.$$

To find the value of $(1.05)^{25}$—

$$\log 1.05 = 0.0211893,$$
$$\log (1.05)^{25} = 25 \times \log 1.05 = 0.5297325,$$
$$\therefore (1.05)^{25} = 3.3864,$$
$$\text{and } S = 40 \times (1.05)^{25} = 135.4560 = £135 \ 9s. \ 2d.$$

If, instead of being payable yearly, the interest had been payable *quarterly*, at the nominal rate of 5 per cent. per annum, the formula would have been

$$S - P\left(1 + \frac{i}{m}\right)^{mn} = 40\left(1 + \frac{.05}{4}\right)^{4 \times 25}$$
$$= 40 \times (1.0125)^{100}.$$

To find the value of 1.0125^{100}, we have

$$\log 1.0125 = 0.0053950,$$
$$\log (1.0125)^{100} = 100 \times \log 1.0125 = 0.539500,$$
$$\therefore (1.0125)^{100} = 3.4634,$$
$$\text{and } S = 40 \times (1.0125)^{100} = 138.536 = £138 \ 10s. \ 8d.$$

(3) The rate of interest being 5 per cent., payable half-yearly, find the sum which, in two years, will amount to £500, allowing for Income Tax at the rate of 6d. per £.

In this case P has to be found from the formula $P = \dfrac{S}{\left(1 + \dfrac{i}{m}\right)^{mn}}$,

where $S = 500$, $i = {\cdot}05$, $m = 2$, and $n = 2$, and Income Tax has to be allowed for at 6d. per £ (*i.e.*, $\cdot025$ per 1).

Accordingly we have

$$P = \frac{500}{\left\{1 + (1 - {\cdot}025) \times \dfrac{{\cdot}05}{2}\right\}^{2 \times 2}},$$

$$= \frac{500}{(1 + {\cdot}975 \times {\cdot}025)^4}$$

$$= \frac{500}{1{\cdot}024375^4} = \frac{500}{1{\cdot}1011},$$

$$= 450{\cdot}08 = \text{£}454 \ 1s. \ 7d.$$

(4) In how many years will a sum of money double itself at compound interest?

To answer this question we use the formula

$$n = \frac{\log S - \log P}{\log (1 + i)}.$$

We have to ascertain how soon 1 will become 2, and, for convenience, taking Napierian logarithms, by which $\log (1 + i)$ is approximately equal to i, we have

$$n = \frac{\log_e 2 - \log_e 1}{\log_e (1 + i)} = \frac{\log_e 2 - \log_e 1}{i}, \text{ approximately.}$$

Now the Napierian logarithm of 2 is $\cdot69$, and of course the logarithm of 1 is 0, so that

$$n = \frac{\cdot69}{i}.$$

From this is derived the common rule that, at compound interest, a sum of money will double itself in $\dfrac{69}{\text{rate per cent.}}$ years. $\longleftarrow$

If a more accurate answer to this question be desired, a result correct to two decimal places will be obtained by estimating that a sum of money will double itself in $\left(\dfrac{69{\cdot}3}{\text{rate per cent.}} + {\cdot}35\right)$ years.

CHAPTER II.

ANNUITIES-CERTAIN.

AMOUNTS AND PRESENT VALUES.
ANNUITY PAYABLE AND INTEREST CONVERTIBLE YEARLY.

Annuities-Certain.

ANNUITIES-CERTAIN are annuities which are payable over a definite period. They are so called to distinguish them from annuities which are dependent for their duration on some contingency—such as the existence of a particular life.

An annuity is measured by the amount payable in the course of a year, which is sometimes called the *rent* of the annuity.

It is assumed (unless otherwise stated) that the first payment of an annuity is due at the end of the year, or, if payable at more frequent intervals, at the end of the first interval. If the first payment be due, not at the end, but at the beginning of the period, the annuity is called an *annuity-due*.

There are two ways in which an annuity-certain may be capitalized, the results being known as the Amount and Present Value of the annuity.

The *amount* is the sum which would represent the total value of the annuity, with interest accrued, on the date when the last payment is made. The *present value*, on the other hand, is the sum which would require to be set aside when the annuity is first entered upon, to meet the payments as they fall due.

The *amount* is the equivalent value of the annuity capitalized at the date when it expires, the *present value* is the capitalized value at the date when the annuity is entered upon,

The present value is accordingly the price that can be paid for the purchase of an annuity, and in investigations connected with annuities-certain, *present value* and *purchase money* are used as synonymous terms.

According to the approved system of notation, the amount of an annuity-certain of 1 for n years is represented by the symbol $s_{\overline{n}|}$, and the present value of a similar annuity, by the symbol $a_{\overline{n}|}$.

Amount at Simple Interest. Simple interest is seldom used in the calculation of annuities, as it leads to very unsatisfactory results. To make our investigations complete, however, we shall begin by considering the question—What is the amount of an annuity of 1 for n years at *Simple Interest?*

To answer this question we take the annuity, payment by payment, beginning with the last. The amount of this payment is of course 1. The payment made one year previously, with a year's interest accrued, amounts to $(1+i)$. The payment made two years previously has two years' interest added, and amounts accordingly to $(1+2i)$. We thus work backwards, each year's payment having an additional year's interest added, till we come to the first payment. Now it is exactly n years since the annuity was entered upon, but the first payment was not made till the end of the first year. It is therefore $(n-1)$ years since the first payment was made, and there are $(n-1)$ years' interest to be added to it. Its amount is accordingly $(1+\overline{n-1}i)$. Adding the amounts of the several payments together, we have, therefore, as the amount of the annuity,

$$s_{\overline{n}|} = 1 + (1+i) + (1+2i) + (1+3i) + \ .\quad . + (1+\overline{n-2}i) + (1+\overline{n-1}i).$$

To find the sum of this series (which is in arithmetical progression), we write it in reverse order—

$$s_{\overline{n}|} = (1+\overline{n-1}i) + (1+\overline{n-2}i) + \quad . . + (1+3i) + (1+2i) + (1+i) + 1.$$

Adding these, term by term—

$$2s_{\overline{n}|} = (2+\overline{n-1}i) + (2+\overline{n-1}i) + (2+\overline{n-1}i) + . . . + (2+\overline{n-1}i),$$
$$= n(2+\overline{n-1}i),$$
$$\therefore s_{\overline{n}|} = \frac{n}{2}(2+\overline{n-1}i),$$

which is the formula for the amount of an annuity-certain of 1 for n years at *simple interest.*

What now is the present value of a similar annuity?

Present Value at Simple Interest. Taking it payment by payment, the present value of the annuity is made up of the value of 1 payable at the end of the first year, that is, $\frac{1}{1+i}$, plus the value of 1 payable at the end of the second year, or $\dfrac{1}{1+2i}$, plus, etc. ; the last term of the series being the present value of 1 payable at the end of n years, or $\dfrac{1}{1+ni}$. We have, therefore,

$$a_{\overline{n}|} = \frac{1}{1+i} + \frac{1}{1+2i} + \frac{1}{1+3i} + \ldots + \frac{1}{1+ni}.$$

This is a series which cannot be summed by any direct method, and it is necessary either to calculate the value of each separate term and add these values together, or to use an approximate method of summation. Several such methods have been suggested, but none of them give satisfactory results, excepting a very complicated formula involving the Integral Calculus.

The present value of an annuity at simple interest is, therefore, very troublesome to calculate, but fortunately it is of little practical importance.

Amount at Compound Interest. Passing now to *Compound Interest*, let us consider what is the amount of a similar annuity of 1, payable yearly for n years.

As before, we begin with the last payment which is just due, and the amount of which is 1. The second last amounts, with interest, to $(1+i)$, and the third last to $(1+i)^2$. We thus work backwards till we come to the first payment which was due $(n-1)$ years ago, and accordingly amounts to $(1+i)^{n-1}$. Adding the amounts of these several payments together, we have as the amount of the annuity—

$$s_{\overline{n}|} = 1 + (1+i) + (1+i)^2 + (1+i)^3 + \ldots + (1+i)^{n-2} + (1+i)^{n-1}.$$

To find the sum of this series (which is in geometrical progression), let us multiply both sides of the equation by $(1+i)$,

$$(1+i)s_{\overline{n}|} = (1+i) + (1+i)^2 + (1+i)^3 + \ldots + (1+i)^{n-1} + (1+i)^n.$$

If now we subtract the original equation from this equation it will be noticed that all the terms on the right hand side cancel each other, excepting $(1+i)^n$ and 1, and we have

$$(1+i)s_{\overline{n}|} - s_{\overline{n}|}, \text{ or, } i s_{\overline{n}|} = (1+i)^n - 1,$$

$$\text{whence, } s_{\overline{n}|} = \frac{(1+i)^n - 1}{i},$$

which is the formula for the amount of an annuity-certain of 1 at *compound interest*.

The same result can be obtained without the use of algebra, as follows :—The amount to which 1 will accumulate in n years at compound interest is $(1+i)^n$, and deducting from this the original capital invested, there remains the amount of the accumulated interest $\left\{(1+i)^n - 1\right\}$. In other words, the amount of an annuity of i for n years is $\left\{(1+i)^n - 1\right\}$. Whence, by simple proportion, the amount of an annuity of 1 is $\dfrac{(1+i)^n - 1}{i}$.

Present Value at Compound Interest.
Let us now consider what is the present value of a similar annuity at *compound interest*.

Taking it payment by payment, as formerly, the value of the payment which is due at the end of the first year is obviously $\dfrac{1}{1+i}$ or v; that of the second payment v^2; of the third v^3, and so on, the value of the last payment being v^n. For the present value of the annuity we, therefore, have the series

$$a_{\overline{n}|} = v + v^2 + v^3 + \ \ldots \ + v^{n-1} + v^n.$$

To find the value of this series (which is in geometrical progression), let us multiply both sides of the equation by v :—

$$va_{\overline{n}|} = v^2 + v^3 + v^4 + \ \ldots \ + v^n + v^{n+1}.$$

Subtract this from the original equation, and it will be noted that all the terms on the right hand side cancel each other, excepting v and v^{n+1}, and we have

$$a_{\overline{n}|} - va_{\overline{n}|} \text{ or } (1-v)a_{\overline{n}|} = v - v^{n+1} = v(1 - v^n),$$

writing for v its value, viz. $\dfrac{1}{1+i}$,

$$\left(1 - \dfrac{1}{1+i}\right)a_{\overline{n}|} = \dfrac{1}{1+i}(1 - v^n),$$

whence, multiplying by $(1+i)$,

$$ia_{\overline{n}|} = 1 - v^n,$$

$$\text{and } a_{\overline{n}|} = \dfrac{1 - v^n}{i}, \quad \longleftarrow$$

which is the formula for the present value of the annuity.

The following is another method of arriving at this result. The present value of an annuity and the present value of its amount should obviously be the same, as it is immaterial whether the annuity be received during n years, or its equivalent value at the end of the time. Accordingly,

$$a_{\overline{n}|} = v^n s_{\overline{n}|},$$

$$= v^n \times \frac{(1+i)^n - 1}{i} = \frac{1}{(1+i)^n} \times \frac{(1+i)^n - 1}{i}$$

$$= \frac{1 - \dfrac{1}{(1+i)^n}}{i} = \frac{1 - v^n}{i}.$$

The formula can also be reasoned out as follows:—

If we invest 1, it will yield an annuity of i for n years, and the present value of the unit which will then be left on hand is, of course, v^n. The amount absorbed in the payment of the annuity of i is therefore $1 - v^n$, and, this being the present value of an annuity of i, the present value of an annuity of 1 is, by simple proportion, $\dfrac{1 - v^n}{i}$.

Amount and Present Value of Annuity-due. An *annuity-due* is, as we have seen, an annuity the first payment of which is made at the time the annuity is entered upon, and not at the end of the first year, as in the case of an ordinary annuity-certain.

Let us briefly consider the amount and present value of such annuities; and first as to the *amount:*—

The last payment having been made at the beginning and not at the end of the last year, one year's interest has to be added to it, and similarly every individual payment will have accumulated for one year longer than in the case of an ordinary annuity. We have, therefore,

$$s = (1+i) + (1+i)^2 + \ldots + (1+i)^n,$$

$$= \left[1 + (1+i) + (1+i)^2 + \ldots + (1+i)^n \right] - 1,$$

$$= s_{\overline{n+1}|} - 1,$$

i.e. the amount of an annuity-due for n years is equal to the amount of an ordinary annuity-certain for $(n+1)$ years, minus 1.

For the *present value* of an annuity-due (the first payment being made at once, and the final payment at the beginning of the last year) we have obviously the series—

$$a_{\overline{n}|} = 1 + v + v^2 + v^3 + \ldots + v^{n-1},$$

$$= 1 + a_{\overline{n-1}|},$$

i.e. the present value of an annuity-due for n years is equal to the present value of an ordinary annuity-certain for $(n-1)$ years, plus 1.

A *Perpetuity* is a perpetual annuity. Feu-duties and
the dividends on consols are familiar examples of
perpetuities.

Value of Perpetuity.

The present value of a perpetuity is easily found by reasoning as
follows :—

A capital sum of 1 will produce a perpetual annuity of i. Therefore, by simple proportion, the sum which must be invested to produce
a perpetuity of 1, or in other words the present value of a perpetuity of 1, is $\frac{1}{i}$, that is $a_\infty = \frac{1}{i}$.

If, instead of being payable at the end of the first year, the first
payment of the perpetuity be due at its commencement, it is obvious
that 1 must be added to the above value. For the present value of
a perpetuity, payable in advance, we therefore have

$$a_\infty = 1 + \frac{1}{i} = \frac{1+i}{i} = \frac{1}{iv} = \frac{1}{d}.$$

Value of Deferred Annuity.

An annuity to run for n years, but which is not to
commence for t years, is called a *Deferred Annuity.*
Let us consider what is the present value of such
an annuity, which is denoted by the symbol $t/a_{\overline{n}|}$.

The first payment will be made at the end of the $(t+1)^{th}$ year,
and its value is accordingly v^{t+1}, and the last payment being made
at the end of the $(t+n)^{th}$ year, its value is similarly v^{t+n}. For the
present value of the deferred annuity we therefore have the series

$$t/a_{\overline{n}|} = v^{t+1} + v^{t+2} + v^{t+3} + \ldots + v^{t+n},$$
$$= v^t\{v + v^2 + v^3 + \ldots + v^n\},$$
$$= v^t a_{\overline{n}|}.$$

Another formula is found as follows :—

$$t/a_{\overline{n}|} = v^{t+1} + v^{t+2} + v^{t+3} + \ldots + v^{t+n},$$
$$= (v + v^2 + \ldots + v^{t+n}) - (v + v^2 + \ldots + v^t),$$
$$= a_{\overline{t+n}|} - a_{\overline{t}|}.$$

We thus see that the value of an annuity-certain for n years,
deferred t years, may be found either (1) by taking the value of an
annuity for n years, and multiplying it by v^t, or (2) by calculating
the value of an annuity for $(t+n)$ years, and subtracting from it the
value of an annuity for t years   The latter formula is usually preferred.

In a similar way it might be shown that the value of a perpetuity
deferred t years—i.e. t/a_∞—can be found either (1) by multiplying
the value of an immediate perpetuity by v^t, or (2) by deducting from

the value of an immediate perpetuity the value of an annuity-certain for t years.

To prevent misunderstanding, it may be well to point out that by an annuity deferred, say 10 years, is not meant an annuity the first payment of which is to be made in 10 years, but an annuity which is not to commence to run for 10 years. The first payment would accordingly be made at the end of the eleventh year.

Construction of Annuity Tables. In forming tables of the amounts and present values of annuities it is advisable, for the reasons given in the last chapter, to work by a continuous process.

Table of Amounts.

It is probable that a table of $(1+i)^n$ will have been formed previously, and if so, the table of $s_{\overline{n}|}$ is constructed very simply from the relation,

$$s_{\overline{n+1}|} = s_{\overline{n}|} + (1+i)^n.$$

Starting with the initial value, $s_{\overline{1}|} = 1$, we have

$$s_{\overline{2}|} = s_{\overline{1}|} + (1+i),$$
$$s_{\overline{3}|} = s_{\overline{2}|} + (1+i)^2,$$
$$s_{\overline{4}|} = s_{\overline{3}|} + (1+i)^3, \text{ and so on.}$$

For verification of the results, the value of $s_{\overline{n}|}$ is calculated, from time to time, by the usual formula $s_{\overline{n}|} = \dfrac{(1+i)^n - 1}{i}$.

If, however, the table of $(1+i)^n$ be not available, a formula for continuous working may be obtained as follows :—

$$s_{\overline{n+1}|} = 1 + (1+i) + (1+i)^2 + \ldots + (1+i)^{n-1} + (1+i)^n,$$
$$= 1 + (1+i)\{1 + (1+i) + \ldots + (1+i)^{n-1}\},$$
$$= 1 + (1+i)s_{\overline{n}|}.$$

Starting, as formerly, with the initial value, we multiply it by $(1+i)$, and adding 1 to the product, obtain the second value. By repeating this process we form the table.

Table of Present Values.

The table of present values can be formed without difficulty if a table of v^n has been previously constructed.

Thus
$$a_{\overline{1}|} = v,$$
$$a_{\overline{2}|} = a_{\overline{1}|} + v^2,$$
$$a_{\overline{3}|} = a_{\overline{2}|} + v^3,$$
$$a_{\overline{4}|} = a_{\overline{3}|} + v^4, \text{ and so on.}$$

To check the results, the value of $a_{\overline{n}|}$ is calculated periodically by the usual formula $a_{\overline{n}|} = \dfrac{1 - v^n}{i}$.

If, however, a table of v^n be not obtainable, we have

$$a_{\overline{n}|} = v + v^2 + v^3 + \ldots + v^n;$$

and, multiplying by $(1+i)$,

$$(1+i)a_{\overline{n}|} = 1 + v + v^2 + \ldots + v^{n-1},$$

$$= 1 + a_{\overline{n-1}|},$$

whence,

$$a_{\overline{n-1}|} = (1+i)a_{\overline{n}|} - 1.$$

To use this formula we first calculate the last value of $a_{\overline{n}|}$ which it is desired to tabulate, and starting with this as the initial value we work backwards. Multiplying the value of $a_{\overline{n}|}$ by $(1+i)$ and deducting 1 we obtain $a_{n-1|}$, and repeating this process we form the table.

Practical Examples.

(1) What, at simple interest, is the amount of an annuity of £40 for 8 years at 5 per cent. ?

The formula we found for the amount of an annuity of 1 at simple interest was

$$s_{\overline{n}|} = \frac{n}{2}\left\{ 2 + (n-1)i \right\}.$$

Substituting the figures given, we have as the amount of an annuity of 1

$$\frac{8}{2}\left\{ 2 + (8-1) \times \cdot05 \right\},$$

$$= 4(2\cdot35) = 9\cdot4.$$

The amount of an annuity of £40 is therefore $40 \times 9\cdot4$ or £376.

(2) What is the present value of this annuity ?

For the present value of an annuity, at simple interest, there is, as we have seen, no convenient formula, and the simplest way of finding it would be to add

$$\frac{1}{1+i} + \frac{1}{1+2i} + \frac{1}{1+3i} + \ldots + \frac{1}{1+ni},$$

in this case

$$\frac{1}{1\cdot05} + \frac{1}{1\cdot10} + \frac{1}{1\cdot15} + \frac{1}{1\cdot20} + \frac{1}{1\cdot25} + \frac{1}{1\cdot30} + \frac{1}{1\cdot35} + \frac{1}{1\cdot40},$$

and multiply the sum by 40.

(3) What would be the amount of this annuity at compound interest ?

The formula in this case is

$$s_{\overline{n}|} = \frac{(1+i)^n - 1}{i}.$$

For the amount of the annuity we therefore have

$$\frac{(1 \cdot 05)^8 - 1}{\cdot 05}.$$

We have first to find the value of $(1 \cdot 05)^8$, and the simplest way to do this is manifestly by logarithms.

Proceeding, however, by ordinary multiplication :—

$$1 \cdot 05 \times 1 \cdot 05 = 1 \cdot 1025 = (1 \cdot 05)^2,$$
$$1 \cdot 1025 \times 1 \cdot 1025 = 1 \cdot 2155 = (1 \cdot 05)^4,$$
$$1 \cdot 2155 \times 1 \cdot 2155 = 1 \cdot 4774 = (1 \cdot 05)^8.$$

We thus have

$$(1 \cdot 05)^8 - 1 = \cdot 4774,$$
$$\text{and,} \quad \frac{(1 \cdot 05)^8 - 1}{\cdot 05} = 9 \cdot 55.$$

This being the amount of an annuity of 1, the amount of an annuity of £40 is, of course, 40 × 9·55 or £382.

(4) What, at compound interest, is the present value of this annuity ?

Here the formula is $a_{\overline{n}|} = \dfrac{1 - v^n}{i}$, or, substituting the given figures

$$a_{\overline{8}|} = \frac{1 - \dfrac{1}{(1 \cdot 05)^8}}{\cdot 05},$$

$$= \frac{1 - \dfrac{1}{1 \cdot 4774}}{\cdot 05} = \frac{1 - \cdot 6768}{\cdot 05} = \frac{\cdot 3232}{\cdot 05},$$

$$= 6 \cdot 464.$$

The present value of an annuity of £40 is accordingly

$$40 \times 6 \cdot 464 = £258 \cdot 56 = £258 \; 11s. \; 2d.$$

We have seen that the present value of an annuity is the same as the present value of its amount.

The amount of this annuity being £382, the present value of the amount is

$$£382 \times v^8 = 382 \times {\cdot}6768 = £258{\cdot}5,$$

which corresponds with the present value of the annuity, as found above.

(5) What is the value, at 5 per cent. interest, of a perpetuity of £40, payable in advance?

Here the formula is $a_\infty = 1 + \dfrac{1}{i}$, and the value of the given perpetuity is accordingly

$$40\left(1 + \frac{1}{{\cdot}05}\right) = 40 \times 21 = £840.$$

(6) What is the value, at 5 per cent. interest, of an annuity-certain of £40 for 8 years, deferred 10 years?

For the value of a deferred annuity there are two formulae,

$$_{10}/a_{\overline{8}|} = v^{10} \times a_{\overline{8}|}, \text{ and}$$
$$_{10}/a_{\overline{8}|} = a_{\overline{18}|} - a_{\overline{10}|},$$

and we can choose whichever is the more convenient.

Taking the first formula, we have $v^{10} = \dfrac{1}{(1{\cdot}05)^{10}} = {\cdot}6139$, and $40 \times a_{\overline{8}|} = 258{\cdot}5$ (as found above).

For the value of the deferred annuity we therefore have

$$258{\cdot}5 \times {\cdot}6139 = £158{\cdot}68 = £158 \ 13s. \ 7d.$$

Taking now the second formula, we have $a_{\overline{18}|} = 11{\cdot}689$, and $a_{\overline{10}|} = 7{\cdot}722$.

For the value of the deferred annuity we therefore have

$$40(11{\cdot}689 - 7{\cdot}722) = £158{\cdot}68 = £158 \ 13s. \ 7d.$$

which agrees with the result already obtained.

CHAPTER III.

ANNUITIES-CERTAIN.

*AMOUNTS AND PRESENT VALUES
WHEN ANNUITY PAYABLE AND INTEREST CONVERTIBLE
AT PERIODS LESS THAN A YEAR.*

In the last chapter we investigated the amounts and present values of annuities when the interest accrued and the annuities were payable once a year.

For the amount of such annuities we found the formula

$$s_{\overline{n}|} = \frac{(1+i)^n - 1}{i},$$

and for their present value

$$a_{\overline{n}|} = \frac{1 - v^n}{i}.$$

We have now to consider the amounts and present values of annuities when the payments are made and the interest is convertible at more frequent intervals. In so doing, we shall first take the cases where the annuities are payable and the interest convertible at the same time—such, for example, as an annuity payable half-yearly, and interest likewise accruing half-yearly; and, secondly, we shall consider the cases where the annuity is payable and the interest convertible at different times—such, for example, as an annuity payable quarterly with interest accruing half-yearly.

Amount of Annuity when Interest convertible and Annuity payable m times a year. What, then, is the amount of an annuity of 1 for n years at compound interest, the annuity being payable and the interest convertible m times a year ?

It is evident that different answers must be given to this question, according as the rate of interest is *nominal* or *effective*. To obviate all confusion in regard to this we shall call the nominal rate j, and the effective rate i.

Taking first the *nominal* rate j, we proceed exactly in the same way as formerly. The sum payable in a year being 1, each payment of the annuity will be $\dfrac{1}{m}$. The amount of the last payment is thus $\dfrac{1}{m}$; the second last (having interest for an mth part of a year added) amounts to $\dfrac{1}{m}\left(1+\dfrac{j}{m}\right)$; the third last to $\dfrac{1}{m}\left(1+\dfrac{j}{m}\right)^2$, and so on, the amount of the first payment being $\dfrac{1}{m}\left(1+\dfrac{j}{m}\right)^{mn-1}$.

We therefore have

$$s = \frac{1}{m} + \frac{1}{m}\left(1+\frac{j}{m}\right) + \frac{1}{m}\left(1+\frac{j}{m}\right)^2 + \ldots + \frac{1}{m}\left(1+\frac{j}{m}\right)^{mn-1},$$

$$= \frac{1}{m}\left\{1 + \left(1+\frac{j}{m}\right) + \left(1+\frac{j}{m}\right)^2 + \ldots + \left(1+\frac{j}{m}\right)^{mn-1}\right\}.$$

Now the series within the large brackets is in geometrical progression, and precisely similar to that for the amount of an annuity payable yearly. By the same process its sum can be shown to be

$$\frac{\left(1+\dfrac{j}{m}\right)^{mn} - 1}{\dfrac{j}{m}}.$$

Therefore,

$$s = \frac{1}{m}\left\{\frac{\left(1+\dfrac{j}{m}\right)^{mn} - 1}{\dfrac{j}{m}}\right\},$$

$$= \frac{\left(1+\dfrac{j}{m}\right)^{mn} - 1}{j}.$$

If, instead of dealing with the nominal rate j, we take the *effective* rate i, the amount of the last payment is, as before, $\dfrac{1}{m}$; the second last amounts to $\dfrac{1}{m}(1+i)^{\frac{1}{m}}$; the third last to $\dfrac{1}{m}(1+i)^{\frac{2}{m}}$, and so on, the first payment amounting to $\dfrac{1}{m}(1+i)^{\frac{mn-1}{m}}$.

We therefore have

$$s = \frac{1}{m} + \frac{1}{m}(1+i)^{\frac{1}{m}} + \frac{1}{m}(1+i)^{\frac{2}{m}} + \ \ldots \ + \frac{1}{m}(1+i)^{\frac{mn-1}{m}},$$

$$= \frac{1}{m}\left\{ 1 + (1+i)^{\frac{1}{m}} + (1+i)^{\frac{2}{m}} + \ \ldots \ + (1+i)^{\frac{mn-1}{m}} \right\}.$$

To find the sum of the series within the large brackets, let

$$\Sigma = 1 + (1+i)^{\frac{1}{m}} + (1+i)^{\frac{2}{m}} + \ \ldots \ + (1+i)^{\frac{mn-1}{m}}.$$

Multiply by $(1+i)^{\frac{1}{m}}$,

$$(1+i)^{\frac{1}{m}}\Sigma = (1+i)^{\frac{1}{m}} + (1+i)^{\frac{2}{m}} + (1+i)^{\frac{3}{m}} + \ \ldots \ + (1+i)^{\frac{mn}{m}}.$$

Subtract from this the original equation, and it will be found that all the terms on the right hand side cancel each other, excepting $(1+i)^{\frac{mn}{m}}$ (which of course is $(1+i)^n$), and 1 ; and we have

$$\left\{ (1+i)^{\frac{1}{m}} - 1 \right\}\Sigma = (1+i)^n - 1,$$

$$\text{or } \Sigma = \frac{(1+i)^n - 1}{(1+i)^{\frac{1}{m}} - 1}.$$

For the amount of the annuity we have accordingly

$$s = \frac{1}{m}\left\{ \frac{(1+i)^n - 1}{(1+i)^{\frac{1}{m}} - 1} \right\} = \frac{(1+i)^n - 1}{m\left\{ (1+i)^{\frac{1}{m}} - 1 \right\}}.$$

Present Value of Annuity when Interest convertible and Annuity payable m times a year. Let us now consider the present value of a similar annuity, taking first the *nominal* rate of interest *j*.

The present value of the first payment of $\frac{1}{m}$, to be made at the end of the first m^{th} part of a year, is

$$\frac{1}{m} \times \frac{1}{\left(1 + \frac{j}{m}\right)} \text{ or } \frac{1}{m}\left(1 + \frac{j}{m}\right)^{-1} ;$$ the value of the second payment is

$$\frac{1}{m} \times \frac{1}{\left(1 + \frac{j}{m}\right)^2} \text{ or } \frac{1}{m}\left(1 + \frac{j}{m}\right)^{-2} ;$$ of the third payment $\frac{1}{m}\left(1 + \frac{j}{m}\right)^{-3}$,

and so on, the value of the last payment being $\frac{1}{m}\left(1 + \frac{j}{m}\right)^{-mn}$.

For the value of the annuity we therefore have

$$a = \frac{1}{m}\left\{\left(1+\frac{j}{m}\right)^{-1} + \left(1+\frac{j}{m}\right)^{-2} + \ \ldots \ + \left(1+\frac{j}{m}\right)^{-mn}\right\}.$$

To find the sum of this series, let

$$\Sigma = \left(1+\frac{j}{m}\right)^{-1} + \left(1+\frac{j}{m}\right)^{-2} + \ \ldots \ + \left(1+\frac{j}{m}\right)^{-mn}.$$

Multiplying by $\left(1+\frac{j}{m}\right)$

$$\left(1+\frac{j}{m}\right)\Sigma = 1 + \left(1+\frac{j}{m}\right)^{-1} + \left(1+\frac{j}{m}\right)^{-2} + \ \ldots \ + \left(1+\frac{j}{m}\right)^{-(mn-1)}.$$

Subtracting now the first equation from the second, all the terms on the right hand side excepting 1 and $\left(1+\frac{j}{m}\right)^{-mn}$ disappear, and we have

$$\left\{\left(1+\frac{j}{m}\right) - 1\right\}\Sigma = 1 - \left(1+\frac{j}{m}\right)^{-mn},$$

$$\therefore \ \Sigma = \frac{1 - \left(1+\frac{j}{m}\right)^{-mn}}{\dfrac{j}{m}},$$

and, accordingly,

$$a = \frac{1}{m}\left\{\frac{1 - \left(1+\frac{j}{m}\right)^{-mn}}{\dfrac{j}{m}}\right\} = \frac{1 - \left(1+\frac{j}{m}\right)^{-mn}}{j}.$$

If, instead of the nominal rate of interest j, we take the *effective* rate i, the present value of the first payment of the annuity is $\dfrac{1}{m}\left\{\dfrac{1}{(1+i)^{\frac{1}{m}}}\right\}$ or $\dfrac{1}{m}(1+i)^{-\frac{1}{m}}$; the value of the second payment is $\dfrac{1}{m}(1+i)^{-\frac{2}{m}}$, and that of the last payment $\dfrac{1}{m}(1+i)^{\frac{-mn}{m}}$ or $\dfrac{1}{m}(1+i)^{-n}$.

Accordingly, for the value of the annuity we have

$$a = \frac{1}{m}\left\{(1+i)^{-\frac{1}{m}} + (1+i)^{-\frac{2}{m}} + \ \ldots \ + (1+i)^{-n}\right\}.$$

This series is in geometrical progression, and is similar to those we have already summed. It amounts to $\dfrac{1 - (1+i)^{-n}}{(1+i)^{\frac{1}{m}} - 1}$.

We have, therefore,

$$a = \frac{1 - (1+i)^{-n}}{m\{(1+i)^{\frac{1}{m}} - 1\}}.$$

Connection between formulae for Annuities payable m times and Annuities payable yearly. There is a close connection between this result and the expression previously found for the value of an annuity payable yearly.

Writing for the value of an annuity-certain for n years, payable m times a year, $a^{(m)}_{\overline{n}|}$,

$$a^{(m)}_{\overline{n}|} = \frac{1 - (1+i)^{-n}}{m\{(1+i)^{\frac{1}{m}} - 1\}} = \frac{1 - (1+i)^{-n}}{i} \times \frac{i}{m\{(1+i)^{\frac{1}{m}} - 1\}}$$

$$= a_{\overline{n}|} \times \frac{i}{m\{(1+i)^{\frac{1}{m}} - 1\}}.$$

But $m\{(1+i)^{\frac{1}{m}} - 1\}$ is the nominal rate of interest corresponding to the effective rate i, and calling this nominal rate j, we have

$$a^{(m)}_{\overline{n}|} = a_{\overline{n}|} \times \frac{i}{j},$$

i.e. the value of an annuity payable m times a year is equal to the value of an annuity payable yearly, multiplied by the effective rate of interest and divided by the corresponding nominal rate.

Similarly, it might be shown that

$$s^{(m)}_{\overline{n}|} = s_{\overline{n}|} \times \frac{i}{j}.$$

We shall now briefly consider the case of an annuity when the payments are made and the interest accrues at different times.

Amount of Annuity when Interest convertible m times and Annuity payable p times. To find a general formula for the amount of this annuity, let us suppose that it is payable p times in the year, and that the interest is convertible m times. We take first the nominal rate of interest j. It is unnecessary to repeat the whole process of reasoning, which is precisely similar to that already employed. We have to bear in mind that the amount to which 1 will accumulate in a year is $\left(1 + \frac{j}{m}\right)^m$, and accordingly in the pth part of a year (which is the interval that elapses between each payment of the annuity) 1 will amount to $\left(1 + \frac{j}{m}\right)^{\frac{m}{p}}.$

As a result we have

which we sum as formerly, and obtain the formula

$$s = \frac{1}{p} \left\{ \frac{\left(1+\frac{j}{m}\right)^{mn} - 1}{\left(1+\frac{j}{m}\right)^{\frac{m}{p}} - 1} \right\}.$$

If, instead of the nominal rate j, the effective rate i be taken we have

$$s = \frac{1}{p} \left\{ 1 + (1+i)^{\frac{1}{p}} + (1+i)^{\frac{2}{p}} + \ldots + (1+i)^{n-\frac{1}{p}} \right\},$$

$$= \frac{1}{p} \left\{ \frac{(1+i)^n - 1}{(1+i)^{\frac{1}{p}} - 1} \right\}.$$

Present value of Annuity when Interest convertible *m* times and Annuity payable *p* times. For the present value of the above annuity, by reasoning similar to that formerly employed, and bearing in mind that, as the present value of 1 due a year hence is $\left(1+\frac{j}{m}\right)^{-m}$, the present value of 1 due in the pth part of a year is $\left(1+\frac{j}{m}\right)^{-\frac{m}{p}}$, we have

$$a = \frac{1}{p} \left\{ \left(1+\frac{j}{m}\right)^{-\frac{m}{p}} + \left(1+\frac{j}{m}\right)^{-\frac{2m}{p}} + \ldots + \left(1+\frac{j}{m}\right)^{-\frac{mpn}{p}} \right\},$$

$$= \frac{1}{p} \left\{ \frac{1 - \left(1+\frac{j}{m}\right)^{-mn}}{\left(1+\frac{j}{m}\right)^{\frac{m}{p}} - 1} \right\}.$$

If, instead of the nominal rate j, we take the effective rate i,

$$a = \frac{1}{p} \left\{ (1+i)^{-\frac{1}{p}} + (1+i)^{-\frac{2}{p}} + \ldots + (1+i)^{-\frac{np}{p}} \right\},$$

$$= \frac{1 - (1+i)^{-n}}{p \left\{ (1+i)^{\frac{1}{p}} - 1 \right\}}.$$

Continuous Annuity. A *Continuous Annuity* is one in which the intervals between the payments are infinitely small. The consideration of such annuities involves the higher mathematics, and is beyond the scope of an elementary treatise.

Tables of Amounts and Present Values. The following Tables show the amounts and present values of an annuity of 1 for n years, payable and with interest convertible yearly, half-yearly, quarterly, and m times a year at the effective rate i, and the nominal rate j. The half-yearly and quarterly results are obtained by adapting the above general formulae.

Annuity Payable	Interest		
	YEARLY.	HALF-YEARLY.	
	Nom. = Effec.	Nominal.	Effective.
Yearly, -	$\dfrac{(1+i)^n - 1}{i}$	$\dfrac{\left(1+\frac{j}{2}\right)^{2n} - 1}{\left(1+\frac{j}{2}\right)^{2} - 1}$	$\dfrac{(1+i)^n - 1}{i}$
Half-Yearly, -	$\dfrac{1}{2}\left\{\dfrac{(1+i)^n - 1}{(1+i)^{\frac{1}{2}} - 1}\right\}$	$\dfrac{\left(1+\frac{j}{2}\right)^{2n} - 1}{j}$	$\dfrac{1}{2}\left\{\dfrac{(1+i)^n - 1}{(1+i)^{\frac{1}{2}} - 1}\right\}$
Quarterly, -	$\dfrac{1}{4}\left\{\dfrac{(1+i)^n - 1}{(1+i)^{\frac{1}{4}} - 1}\right\}$	$\dfrac{1}{4}\left\{\dfrac{\left(1+\frac{j}{2}\right)^{2n} - 1}{\left(1+\frac{j}{2}\right)^{\frac{1}{2}} - 1}\right\}$	$\dfrac{1}{4}\left\{\dfrac{(1+i)^n - 1}{(1+i)^{\frac{1}{4}} - 1}\right\}$
m times a year,	$\dfrac{1}{m}\left\{\dfrac{(1+i)^n - 1}{(1+i)^{\frac{1}{m}} - 1}\right\}$	$\dfrac{1}{m}\left\{\dfrac{\left(1+\frac{j}{2}\right)^{2n} - 1}{\left(1+\frac{j}{2}\right)^{\frac{2}{m}} - 1}\right\}$	$\dfrac{1}{m}\left\{\dfrac{(1+i)^n - 1}{(1+i)^{\frac{1}{m}} - 1}\right\}$

Present Value of Annuity

Annuity Payable	Interest		
	YEARLY.	HALF-YEARLY.	
	Nom. = Effec.	Nominal.	Effective.
Yearly, -	$\dfrac{1 - (1+i)^{-n}}{i}$	$\dfrac{1 - \left(1+\frac{j}{2}\right)^{-2n}}{\left(1+\frac{j}{2}\right)^{2} - 1}$	$\dfrac{1 - (1+i)^{-n}}{i}$
Half-Yearly, -	$\dfrac{1}{2}\left\{\dfrac{1 - (1+i)^{-n}}{(1+i)^{\frac{1}{2}} - 1}\right\}$	$\dfrac{1 - \left(1+\frac{j}{2}\right)^{-2n}}{j}$	$\dfrac{1}{2}\left\{\dfrac{1 - (1+i)^{-n}}{(1+i)^{\frac{1}{2}} - 1}\right\}$
Quarterly, -	$\dfrac{1}{4}\left\{\dfrac{1 - (1+i)^{-n}}{(1+i)^{\frac{1}{4}} - 1}\right\}$	$\dfrac{1}{4}\left\{\dfrac{1 - \left(1+\frac{j}{2}\right)^{-2n}}{\left(1+\frac{j}{2}\right)^{\frac{1}{2}} - 1}\right\}$	$\dfrac{1}{4}\left\{\dfrac{1 - (1+i)^{-n}}{(1+i)^{\frac{1}{4}} - 1}\right\}$
m times a year,	$\dfrac{1}{m}\left\{\dfrac{1 - (1+i)^{-n}}{(1+i)^{\frac{1}{m}} - 1}\right\}$	$\dfrac{1}{m}\left\{\dfrac{1 - \left(1+\frac{j}{2}\right)^{-2n}}{\left(1+\frac{j}{2}\right)^{\frac{2}{m}} - 1}\right\}$	$\dfrac{1}{m}\left\{\dfrac{1 - (1+i)^{-n}}{(1+i)^{\frac{1}{m}} - 1}\right\}$

Convertible

QUARTERLY		m TIMES A YEAR	
Nominal.	Effective.	Nominal.	Effective.
$\dfrac{\left(1+\frac{j}{4}\right)^{4n}-1}{\left(1+\frac{j}{4}\right)^{4}-1}$	$\dfrac{(1+i)^n-1}{i}$	$\dfrac{\left(1+\frac{j}{m}\right)^{mn}-1}{\left(1+\frac{j}{m}\right)^{m}-1}$	$\dfrac{(1+i)^n-1}{i}$
$\frac{1}{2}\left\{\dfrac{\left(1+\frac{j}{4}\right)^{4n}-1}{\left(1+\frac{j}{4}\right)^{2}-1}\right\}$	$\frac{1}{2}\left\{\dfrac{(1+i)^n-1}{(1+i)^{\frac12}-1}\right\}$	$\frac{1}{2}\left\{\dfrac{\left(1+\frac{j}{m}\right)^{mn}-1}{\left(1+\frac{j}{m}\right)^{\frac{m}{2}}-1}\right\}$	$\frac{1}{2}\left\{\dfrac{(1+i)^n-1}{(1+i)^{\frac12}-1}\right\}$
$\dfrac{\left(1+\frac{j}{4}\right)^{4n}-1}{j}$	$\frac{1}{4}\left\{\dfrac{(1+i)^n-1}{(1+i)^{\frac14}-1}\right\}$	$\frac{1}{4}\left\{\dfrac{\left(1+\frac{j}{m}\right)^{mn}-1}{\left(1+\frac{j}{m}\right)^{\frac{m}{4}}-1}\right\}$	$\frac{1}{4}\left\{\dfrac{(1+i)^n-1}{(1+i)^{\frac14}-1}\right\}$
$\frac{1}{m}\left\{\dfrac{\left(1+\frac{j}{4}\right)^{4n}-1}{\left(1+\frac{j}{4}\right)^{\frac{4}{m}}-1}\right\}$	$\frac{1}{m}\left\{\dfrac{(1+i)^n-1}{(1+i)^{\frac{1}{m}}-1}\right\}$	$\dfrac{\left(1+\frac{j}{m}\right)^{mn}-1}{j}$	$\frac{1}{m}\left\{\dfrac{(1+i)^n-1}{(1+i)^{\frac{1}{m}}-1}\right\}$

Of 1 for n *Years.*

Convertible

QUARTERLY		m TIMES A YEAR	
Nominal.	Effective.	Nominal.	Effective.
$\dfrac{1-\left(1+\frac{j}{4}\right)^{-4n}}{\left(1+\frac{j}{4}\right)^{4}-1}$	$\dfrac{1-(1+i)^{-n}}{i}$	$\dfrac{1-\left(1+\frac{j}{m}\right)^{-mn}}{\left(1+\frac{j}{m}\right)^{m}-1}$	$\dfrac{1-(1+i)^{-n}}{i}$
$\frac{1}{2}\left\{\dfrac{1-\left(1+\frac{j}{4}\right)^{-4n}}{\left(1+\frac{j}{4}\right)^{2}-1}\right\}$	$\frac{1}{2}\left\{\dfrac{1-(1+i)^{-n}}{(1+i)^{\frac12}-1}\right\}$	$\frac{1}{2}\left\{\dfrac{1-\left(1+\frac{j}{m}\right)^{-mn}}{\left(1+\frac{j}{m}\right)^{\frac{m}{2}}-1}\right\}$	$\frac{1}{2}\left\{\dfrac{1-(1+i)^{-n}}{(1+i)^{\frac12}-1}\right\}$
$\dfrac{1-\left(1+\frac{j}{4}\right)^{-4n}}{j}$	$\frac{1}{4}\left\{\dfrac{1-(1+i)^{-n}}{(1+i)^{\frac14}-1}\right\}$	$\frac{1}{4}\left\{\dfrac{1-\left(1+\frac{j}{m}\right)^{-mn}}{\left(1+\frac{j}{m}\right)^{\frac{m}{4}}-1}\right\}$	$\frac{1}{4}\left\{\dfrac{1-(1+i)^{-n}}{(1+i)^{\frac14}-1}\right\}$
$\frac{1}{m}\left\{\dfrac{1-\left(1+\frac{j}{4}\right)^{-4n}}{\left(1+\frac{j}{4}\right)^{\frac{4}{m}}-1}\right\}$	$\frac{1}{m}\left\{\dfrac{1-(1+i)^{-n}}{(1+i)^{\frac{1}{m}}-1}\right\}$	$\dfrac{1-\left(1+\frac{j}{m}\right)^{-mn}}{j}$	$\frac{1}{m}\left\{\dfrac{1-(1+i)^{-n}}{(1+i)^{\frac{1}{m}}-1}\right\}$

Application of Table to Perpetuities.

The foregoing table for the present value of annuities can be rendered applicable to perpetuities by making the numerator in each case 1.

It will be seen that throughout the table the numerators take the general form $\left(1 - \dfrac{1}{(1+i)^n}\right)$, or when the rate of interest used is nominal $\left\{1 - \dfrac{1}{\left(1+\dfrac{j}{m}\right)^{m n}}\right\}$.

Now, if n—the number of years—be infinitely great, as in the case of perpetuities, the expressions $(1+i)^n$ and $\left(1+\dfrac{j}{m}\right)^{mn}$ become infinitely great likewise, and the fractions in consequence infinitely small. The numerators, accordingly, become in each case 1, minus an infinitely small fraction, *i.e.* 1.

It will be noted that when a perpetuity is payable and the interest convertible at the same time, no matter how frequently in the course of the year, the value is always the same, viz.,

$$\frac{1}{\text{nominal rate of interest.}}$$

Practical Examples.

(1) Find the amount of an annuity of £50 for 6 years, with interest and instalment payable half-yearly, the nominal rate of interest being 5 per cent. ?

The formula in this case is

$$\frac{\left(1 + \dfrac{j}{2}\right)^{2n} - 1}{j};$$

and, substituting the given figures, we have for the amount of the annuity

$$50 \times \frac{\left(1 + \dfrac{\cdot 05}{2}\right)^{12} - 1}{\cdot 05}.$$

To find the value of $(1 \cdot 025)^{12}$,

$$\log (1 \cdot 025)^{12} = 12 \log 1 \cdot 025 = 12 \times 0 \cdot 0107239 = 0 \cdot 1286868,$$
$$\therefore (1 \cdot 025)^{12} = 1 \cdot 34489,$$

whence, $\dfrac{(1 \cdot 025)^{12} - 1}{\cdot 05} = 6 \cdot 8978,$

and, $50 \times \dfrac{(1 \cdot 025)^{12} - 1}{\cdot 05} = 344 \cdot 9 = £344$ 18s.

(2) *If, in the above example, 5 per cent. were the effective rate, what would be the amount of the annuity ?*

The formula here is

$$\frac{1}{2}\left\{\frac{(1+i)^n - 1}{(1+i)^{\frac{1}{2}} - 1}\right\};$$

and, accordingly, we have for the amount of the annuity

$$50 \times \frac{1}{2}\left\{\frac{(1\cdot05)^6 - 1}{(1\cdot05)^{\frac{1}{2}} - 1}\right\} = £344 \; 6s.$$

(3) *What now would be the present value of this annuity at the nominal rate of 5 per cent. ?*

As will be seen from the table, the formula is

$$\frac{1 - \left(1 + \frac{j}{2}\right)^{-2n}}{j};$$

and, substituting the given figures and the value already found for $\left(1 + \frac{j}{2}\right)^{2n}$, we have for the value of the annuity of £50

$$50 \times \frac{1 - \dfrac{1}{1\cdot34489}}{\cdot05} = 1000(1 - \cdot7436),$$

$$= 256\cdot4 = £256 \; 8s.$$

We have seen that the present value of an annuity is the same as the present value of its amount.

The amount of this annuity we found to be 344·9, and the present value of the amount is

$$344\cdot9 \times (1\cdot025)^{-12} = 344\cdot9 \times \cdot7436 = 256\cdot4,$$

which corresponds with the present value of the annuity as found above.

(4) *What is the present value of an annuity of £50 for 6 years, at a nominal rate of interest of 5 per cent., the annuity being payable quarterly, and the interest convertible half-yearly ?*

Here the formula is

$$\frac{1}{p}\left\{\frac{1 - \left(1 + \frac{j}{m}\right)^{-mn}}{\left(1 + \frac{j}{m}\right)^{\frac{m}{p}} - 1}\right\}.$$

Annuities-Certain.

We have, therefore, as the value of the annuity

$$50 \times \frac{1}{4} \left\{ \frac{1-(1 \cdot 025)^{-12}}{(1 \cdot 025)^{\frac{2}{4}}-1} \right\},$$

$$= \frac{50}{4} \times \frac{\cdot 2564}{\cdot 0124} = 258 \cdot 5 = £258 \quad 10s.$$

(5) If an annuity of £845 payable half-yearly be converted into an annuity payable quarterly, the effective rate of interest being $3\frac{3}{4}$ per cent., what sum will be paid per annum instead of £845?

Here we have

$$845 \left\{ \frac{1-(1 \cdot 0375)^{-n}}{2\left\{(1 \cdot 0375)^{\frac{1}{2}}-1\right\}} \right\} = x \left\{ \frac{1-(1 \cdot 0375)^{-n}}{4\left\{(1 \cdot 0375)^{\frac{1}{4}}-1\right\}} \right\},$$

$$\text{whence,} \quad x = 845 \left\{ \frac{4(1 \cdot 0375^{\frac{1}{4}}-1)}{2(1 \cdot 0375^{\frac{1}{2}}-1)} \right\},$$

$$= 845 \times \frac{\cdot 01850}{\cdot 01857},$$

$$= 841 \cdot 8 = £841 \quad 16s.$$

(6) The present value of an annuity of £1, payable yearly for 6 years at 5 per cent. interest, being £5·0757, find the value of an annuity of £1 payable half-yearly, the nominal rate of interest corresponding to an effective rate of 5 per cent. being 4·939.

The formula here is

$$a_{\overline{n}|}^{m} = a_{\overline{n}|} \times \frac{i}{j} \quad (a_{\overline{n}|} \text{ being taken at rate } i);$$

and, substituting the given values

$$a_{\overline{6}|}^{2} = 5 \cdot 0757 \times \frac{\cdot 05}{\cdot 04939},$$

$$= £5 \cdot 1383.$$

CHAPTER IV.

ANNUITIES-CERTAIN.

*APPORTIONMENT OF PAYMENTS BETWEEN
CAPITAL AND INTEREST.
SINKING FUNDS.*

*Investment of
Purchase Money
provides Annual
Payments.*

WE have seen that the present value or purchase money of an annuity of 1, payable yearly for n years, is represented by the series $(v + v^2 + v^3 + \ldots + v^n)$,

or $\dfrac{1}{(1+i)} + \dfrac{1}{(1+i)^2} + \dfrac{1}{(1+i)^3} + \ldots + \dfrac{1}{(1+i)^n}.$

If, now, this purchase money be invested at the rate of interest at which the annuity is calculated, it will amount at the end of the first year to

$$(1+i)\left\{ \dfrac{1}{(1+i)} + \dfrac{1}{(1+i)^2} + \dfrac{1}{(1+i)^3} + \ldots + \dfrac{1}{(1+i)^n} \right\},$$

$$= 1 + \dfrac{1}{(1+i)} + \dfrac{1}{(1+i)^2} + \ldots + \dfrac{1}{(1+i)^{n-1}},$$

$$= 1 + v + v^2 + \ldots + v^{n-1}.$$

The investment of the purchase money thus provides the 1 required to pay the instalment of the annuity due at the end of the first year, and after making this payment, the sum of $(v + v^2 + v^3 + \ldots + v^{n-1})$ remains on hand. Now it will be noted that this sum is equivalent to $a_{\overline{n-1}|}$, so that, after paying the instalment, the value of an annuity for the remaining $(n-1)$ years is still invested.

At the end of the second year this sum, with interest added, will amount to

$$(1+i)\left\{ v + v^2 + \ldots + v^{n-1} \right\},$$

$$= 1 + v + v^2 + \ldots + v^{n-2},$$

which again provides the 1 required, and leaves as invested capital $(v + v^2 + \ldots + v^{n-2})$, which is the value of an annuity for the remaining $(n-2)$ years. If this process be repeated, it is evident that, year by year, the required payment is provided, and a regularly decreasing amount is left on hand, which in each case is the value of an annuity for the remainder of the n years.

We thus see that the granter of the annuity will be regularly provided with the amount required to pay the instalments as they fall due, from the investment of the purchase money which he receives.

Apportionment of Annual Payments between Capital and Interest. Let us now look at the matter from the point of view of the purchaser of the annuity. In consideration of an immediate payment of $(v + v^2 + v^3 + \ldots + v^n)$, or $a_{\overline{n}|}$, he is to receive 1 per annum for n years. This annual instalment has not only to pay interest on the purchase money, but also to provide for the gradual return of the capital. We must therefore consider how the various payments of the annuity are to be apportioned between *capital* and *interest*.

The original value of the annuity is

$$v + v^2 + v^3 + \ldots + v^{n-1} + v^n,$$

and we have seen that after the first payment has been made there still remains in the hands of the granter

$$v + v^2 + v^3 + \ldots + v^{n-2} + v^{n-1},$$

and after the second

$$v + v^3 + v^3 + \ldots + v^{n-2},$$

and that this amount decreases regularly with each succeeding payment. Obviously, therefore, the capital repaid in the first instalment of the annuity is v^n; in the second instalment v^{n-1}; in the third v^{n-2}; and in the last v.

These repayments of capital amount in the aggregate to

$$v^n + v^{n-1} + v^{n-2} + \ldots + v = a_{\overline{n}|},$$

which was the sum originally invested.

The question still remains, whether, after these sums have been applied in repayment of capital, the balance of the instalments will be sufficient to pay interest on the investment.

At the end of the first year the purchaser receives 1, and after setting aside v^n to repay capital, he is left with

$$(1 - v^n) = i \times \frac{1 - v^n}{i} = i a_{\overline{n}|},$$

which is the interest on the sum of $a_{\overline{n}|}$ originally invested.

At the beginning of the second year the amount remaining invested is $a_{\overline{n-1}|}$. At the end of the year the purchaser receives 1 and sets aside v^{n-1}, so that he is left with

$$(1 - v^{n-1}) = i \times \frac{1 - v^{n-1}}{i} = i a_{\overline{n-1}|},$$

which again is the interest on the invested capital.

The same holds good with each succeeding year, and it is therefore evident that, after providing the required repayment of capital, the balance of each instalment of the annuity is exactly sufficient to pay interest on the sum ~~remaining~~ invested.

The following table shows the relative amounts of capital and interest contained in each payment of an annuity of 1 in accordance with the principles above established.

Year.	Amount Invested at Beginning of Year.	Interest.		Capital Repaid.	Capital Outstanding at End of Year.
1	$a_{\overline{n}|}$	$i a_{\overline{n}|}$	or $1 - v^n$	v^n	$a_{\overline{n}|} - v^n = a_{\overline{n-1}|}$
2	$a_{\overline{n-1}|}$	$i a_{\overline{n-1}|}$	or $1 - v^{n-1}$	v^{n-1}	$a_{\overline{n-1}|} - v^{n-1} = a_{\overline{n-2}|}$
3	$a_{\overline{n-2}|}$	$i a_{\overline{n-2}|}$	or $1 - v^{n-2}$	v^{n-2}	$a_{\overline{n-2}|} - v^{n-2} = a_{\overline{n-3}|}$
4	$a_{\overline{n-3}|}$	$i a_{\overline{n-3}|}$	or $1 - v^{n-3}$	v^{n-3}	$a_{\overline{n-3}|} - v^{n-3} = a_{\overline{n-4}|}$
.	.	.		.	.
2nd last	$a_{\overline{2}|}$	$i a_{\overline{2}|}$	or $1 - v^2$	v^2	$a_{\overline{2}|} - v^2 = a_{\overline{1}|}$
last	$a_{\overline{1}|}$	$i a_{\overline{1}|}$	or $1 - v$	v	$a_{\overline{1}|} - v = 0$

Capital and Interest in *m*th payment. It is often necessary to determine the amount of capital and interest respectively contained in a specified payment of an annuity, say the mth. On referring to the above table, it will be seen that the capital contained in the third payment is $v^{n-(3-1)}$, and in the fourth $v^{n-(4-1)}$. Similarly, the capital contained in the mth payment is $v^{n-(m-1)}$, or v^{n-m+1}, and the interest is, of course, the balance of the payment, or $1 - v^{n-m+1}$.

Capital Outstanding. It will be noted likewise that the capital outstanding at the commencement of any year is simply the value of an annuity for the unexpired period.

E

Let us now suppose that, instead of a portion of each annual instalment being applied in repayment of the purchase money, a separate *Sinking Fund* is established to replace the capital on the expiry of the annuity.

The amount invested in the annuity being $a_{\overline{n}|}$, and the annual interest on this investment $ia_{\overline{n}|}$, the balance of each annual payment available for such a *Sinking Fund* (after providing for interest) is

$$S.\,F.\qquad (1 - ia_{\overline{n}|}) = \left(1 - i \times \frac{1 - v^n}{i}\right) = v^n.$$

We have now to show that the yearly investment of this balance at the same rate of interest as the annuity yields will reproduce the original capital; or in other words, that $(1 - ia_{\overline{n}|})\,s_{\overline{n}|}$ is equal to $a_{\overline{n}|}$. We have

$$(1 - ia_{\overline{n}|})\,s_{\overline{n}|} = \left(1 - i \times \frac{1 - v^n}{i}\right)\frac{(1 + i)^n - 1}{i},$$

$$= \frac{1}{(1 + i)^n} \times \frac{(1 + i)^n - 1}{i},$$

$$= \frac{1 - \dfrac{1}{(1 + i)^n}}{i} = a_{\overline{n}|}.$$

It is evident, therefore, that a *Sinking Fund* of $(1 - ia_{\overline{n}|})$ or v^n per annum, accumulated for n years at the same rate of interest as the annuity yields, will exactly amount to the sum originally invested, and we establish the very important relation

$$a_{\overline{n}|} = (1 - ia_{\overline{n}|})\,s_{\overline{n}|}$$

or, transposing this equation so as to express $a_{\overline{n}|}$ in terms of $s_{\overline{n}|}$,

$$a_{\overline{n}|} = \frac{s_{\overline{n}|}}{1 + is_{\overline{n}|}}.$$

If, as sometimes happens, the rate of interest which the annuity is intended to yield be higher than that at which the *Sinking Fund* can be invested, the question becomes more complicated, as two rates of interest have to be taken into account. We have to consider the *remunerative* rate, which the purchaser wishes to realise on his investment, and which we shall call i, and the smaller *accumulative* rate at which the *Sinking Fund* is to be invested, and which we shall call j.

The value of an annuity of this description of 1 for n years, which will pay the purchaser interest on his investment at the rate i, the *Sinking Fund* being accumulated at the smaller rate j, is found from the above formula

$$a_{\overline{n}|} = \frac{s_{\overline{n}|}}{1 + i s_{\overline{n}|}},$$

by calculating $s_{\overline{n}|}$ at the accumulative rate j

If s' be the amount of an annuity of 1 for n years calculated at the rate j, the price paid will thus be $\frac{s'}{1 + i s'}$, and as the interest included in the first payment is $\frac{i s'}{1 + i s'}$, the balance of the instalment available for a *Sinking Fund* (after providing for interest) will be

$$1 - \frac{i s'}{1 + i s'}, \text{ or } \frac{1}{1 + i s'}.$$

If this sum of $\frac{1}{1 + i s'}$ be set aside annually, and accumulated at the rate of interest j, it will amount at the end of n years to $\frac{1}{1 + i s'} \times s'$, or $\frac{s'}{1 + i s'}$, and so reproduce the amount invested in the annuity.

Annuities of this description are usually employed in the valuation of mining rights.

Practical Examples.

(1) What is the capital outstanding after the sixth payment of an annuity-certain of £40, the annuity being calculated at 5 per cent. interest, and its original duration 25 years ?

The capital outstanding is equal to the value of an annuity for the remainder of the period. After the sixth payment, therefore, the capital outstanding is

$$40 \times a_{\overline{19}|} = 40 \times \frac{1 - \frac{1}{(1 \cdot 05)^{19}}}{\cdot 05}$$

$$= 40 \times 12 \cdot 085 = £483 \ 8s.$$

(2) What amounts of capital and interest respectively are contained in the seventh payment of the foregoing annuity?

The capital is

$$40 \times v^{25-7+1} = 40 \times v^{19}$$

$$= 40 \times \cdot3957 = £15 \ 16s. \ 7d.$$

The interest is the balance of the payment,

$$40 - 15 \cdot 83 \quad = 24 \cdot 17 \quad = £24 \ 3s. \ 5d.$$

It will be observed that the interest is exactly 5 per cent. on the capital outstanding after the sixth payment, which, as we have seen, amounts to £483 8s.

(3) What is the Sinking Fund which must be invested annually at 5 per cent. to replace the capital on the expiry of the above annuity?

The rate of interest at which the sinking fund is to be invested is the same as that at which the annuity is calculated. The sinking fund is accordingly

$$\left(1 - i\ a\,\overline{\underset{\frown}{}} \right) \qquad 40 \times (1 - \cdot 05 \ a_{\overline{25}|}) \text{ or } 40 \times v^{25}$$

$$= 40 \times \cdot 2953 = £11 \ 16s. \ 3d.$$

(4) What is the value of an annuity-certain of £40 for 25 years to pay 5 per cent. on the investment, the Sinking Fund being accumulated at 4 per cent.?

The expression we found for the value of an annuity of this description was $\left(\dfrac{s'}{1+is'} \right)$ and, applying this formula we have

$$40 \times \frac{s_{\overline{25}|}{}_{\cdot 4}}{1 + \cdot 05 \ s_{\overline{25}|}{}_{\cdot 4}} \ (s_{\overline{25}|} \text{ being taken at 4 per cent.}),$$

$$= 40 \times \frac{41 \cdot 646}{1 + \cdot 05 \times 41 \cdot 646}$$

$$= 40 \times \frac{41 \cdot 646}{3 \cdot 082} = £540 \ 10s.$$

(5) *The annual profit of a Colliery with a lease having 19 years to run is £5000. What price can be paid by a purchaser so as to realize 12 per cent. on the investment, the Redemption Fund being accumulated at 4 per cent. ?*

This is another example of the application of the same formula, and we accordingly have as the price to be paid

$$5000 \times \frac{s_{\overline{19}|}}{1 + \cdot 12 \ s_{\overline{19}|}} \quad (s_{\overline{19}|} \text{ being taken at 4 per cent.})$$

$$= 5000 \times \frac{27 \cdot 6712}{1 + \cdot 12 \times 27 \cdot 6712}$$

$$= \frac{138356}{4 \cdot 3205} = £32,023 \ \ 3s.$$

(6) *Under a Feu Contract a casualty of £195 7s. is payable every nineteenth year, the first payment being due 9 years hence. It is desired to do away with the casualty by adding an equivalent amount to the annual feu-duty—calculating same at 4 per cent. interest. What annual addition to the feu-duty is required?*

The first step is to ascertain the present value of the casualty, which is a perpetuity of £195·35 with 19 years between the payments. The interest for 19 years, corresponding to 4 per cent. per annum, is $\{(1 \cdot 04)^{19} - 1\}$, and accordingly if the first payment had been due at once the value of the perpetuity (see page 15) would have been

$$195 \cdot 35 \times \left\{ 1 + \frac{1}{(1 \cdot 04)^{19} - 1} \right\}$$

$$= 195 \cdot 35 \times 1 \cdot 9033 = 371 \cdot 810.$$

As, however, the first payment (instead of being made at once) is not due for 9 years, we must multiply this value by v^9. For the present value of the casualty we therefore have

$$371 \cdot 810 \times \cdot 7026 = 261 \cdot 234.$$

To ascertain the yearly addition to the feu-duty it now only remains to determine the perpetuity at 4 per cent. which £261·234 will purchase. This of course is

$$261 \cdot 234 \times \cdot 04 = 10 \cdot 45 = £10 \ \ 9s.$$

CHAPTER V.

ANNUITIES-CERTAIN.

LOANS REPAYABLE BY TERMINABLE ANNUITIES.

Annuity required to repay Loan. To determine the annuity required to repay a loan with interest in a definite number of years, we have simply to apply the formulae already found for the present values of annuities.

An annual payment of 1 for n years will repay, with interest, an advance of $a_{\overline{n}|}$. Accordingly, to find the annuity which will repay an advance say of P, we have

$$a_{\overline{n}|} : P :: 1 : \text{the required annuity.}$$

The terminable annuity, therefore, which will repay with interest a loan of P in n years is $\dfrac{P}{a_{\overline{n}|}}$.

Sinking Fund. To find the amount of the sinking fund, or capital repaid in the first instalment of this annuity, we may proceed in two ways—

(1) As the interest contained in the first payment is iP, the balance of the payment of $\dfrac{P}{a_{\overline{n}|}}$ available for a sinking fund is

$$\left(\frac{P}{a_{\overline{n}|}} - iP\right), \text{ or } P\left(\frac{1}{a_{\overline{n}|}} - i\right).$$

(2) As 1 per annum will amount to $s_{\overline{n}|}$ in n years, the sum which will accumulate to P is, by simple proportion, $\dfrac{P}{s_{\overline{n}|}}$.

That these two expressions for the sinking fund are identical is easily shown, thus :—

$$P\left(\frac{1}{a_{\overline{n}|}} - i\right) = P\left\{\frac{i}{1 - v^n} - i\right\} = P\left\{\frac{iv^n}{1 - v^n}\right\}$$

$$= P\left\{\frac{i}{\frac{1}{v^n} - 1}\right\} = P\left\{\frac{i}{(1+i)^n - 1}\right\}$$

$$= \frac{P}{s_{\overline{n}|}}.$$

Apportionment of Instalments between Capital and Interest. This sinking fund may either be separately invested each year at the same rate of interest as the annuity yields, or a portion of each instalment of the annuity may be taken as capital repaid. In the latter event the amount of each successive repayment is obtained by multiplying by $(1+i)$ the capital contained in the preceding payment, thus

$$\text{Capital contained in 1st payment} = \frac{P}{s_{\overline{n}|}},$$

$$\text{,,} \qquad \text{,,} \qquad \text{2nd} \quad \text{,,} \quad = \frac{P}{s_{\overline{n}|}}(1+i),$$

$$\text{,,} \qquad \text{,,} \qquad \text{3rd} \quad \text{,,} \quad = \frac{P}{s_{\overline{n}|}}(1+i)^2,$$

$$\text{,,} \qquad \text{,,} \qquad \text{4th} \quad \text{,,} \quad = \frac{P}{s_{\overline{n}|}}(1+i)^3,$$

and so on, the interest in each case being, of course, the balance of the payment.

If a table of the amounts of principal and interest contained in each payment of an annuity be constructed in this way, the formulae already found (see page 33) may be used for verification. Thus, in the mth payment the capital will be $\frac{P}{a_{\overline{n}|}} \times v^{n-m+1}$. The correctness of the calculation is also proved by the principal being exactly repaid on the expiry of the annuity.

Annuity required to repay Loan when Sinking Fund accumulated at lower rate than Loan yields. If, as frequently happens, it be a condition of the loan that the sinking fund is to be accumulated at a lower rate of interest than the loan itself is intended to yield, the formula investigated in last chapter is employed.

We found that the value of an annuity of this description of 1 per annum is represented by the expression

$$A = \frac{s'}{1+is'},$$

where i is the remunerative rate of interest, and s' is calculated at the accumulative rate j. The corresponding annuity which will repay a loan of P is, by simple proportion,

$$\frac{P}{A} = \frac{P(1+is')}{s'} = P\left\{\frac{1}{s'}+i\right\},$$

and, writing for $\frac{1}{s'}$ its equivalent $\left(\frac{1}{a'}-j\right)$, we have

$$\frac{P}{A}=P\left\{\frac{1}{a'}+(i-j)\right\}.$$

To find the terminable annuity therefore, which will repay a loan of P with interest at the rate i, the sinking fund being accumulated at the rate j, we calculate (or take from the table) the annuity which 1 will purchase at rate j $\left(i.e.\ \frac{1}{a'}\right)$, add to this the difference in the rates of interest $(i-j)$, and multiply the sum by P.

Loans issued at a discount and repayable at par, etc. In connection with loans, whether repayable in one sum, or by instalments over a period of years, many complications arise. The loan, for example, may be issued at par and repayable with a bonus, or issued at a discount and repayable at par.

The method which naturally suggests itself for dealing with loans of this description is to separate them carefully into their component parts and value each of these parts by itself. Suppose for instance (to take a very simple example) that the loan (£1000) is to run for 5 years at 4 per cent. interest, and that the principal is repayable at the end of the period with a bonus of 5 per cent. What would be the value of this bond to a purchaser who wished to realise say 3 per cent. on his investment?

Here we have (1) a principal sum of £1050 payable at the end of 5 years, the present value of which at 3 per cent. is $1050 \times v^5$, and (2) an annuity of £40 for 5 years, the value of which is $40\ a_{\overline{5}|}$ taken at 3 per cent. The sum of these values is the value of the bond.

Makeham's Formula. The foregoing is a method which is always applicable. It is frequently laborious, however, and in such cases as that above cited, in which the sums receivable in name of interest bear always the same proportion to the capital at the time outstanding, the following general formula devised by the late W. M. Makeham offers great economy of labour.

To arrive at this general formula, let

$A =$ the value of the loan to an investor.
$C =$ the capital sum repayable, *including any bonus.*
$K =$ the present value of C at rate i.
$i =$ the rate of interest which it is desired to realise on the investment.

$j =$ the rate of interest on C which is represented by the annual interest payable.

Thus if the loan be for £1000 at 3 per cent., repayable at maturity with a bonus of 25 per cent., C is £1250, and j (it must be specially noted) is not ·03, but $\frac{1000}{1250} \times$ ·03, or ·024.

To ascertain the value of the loan, we must consider separately (1) the value of the capital sum repayable, which is denoted by K, and (2) the value at rate i of the annual payment of interest. Taken together these make up the value of the loan, *i.e.* A.

Now if the loan had been returning interest at the rate i the value of the whole loan would have been C, and the value of the annual interest—iC—would have been $(C - K)$.

It follows, therefore, that the value of each unit of interest is $\frac{C - K}{iC}$.

By hypothesis, however, the actual interest realised annually is jC, and as the value of each unit is $\frac{C - K}{iC}$, the value of this interest will be $jC \times \frac{C - K}{iC}$, or $\frac{j}{i}(C - K)$.

For the total value of the loan we have accordingly (1) the value of the principal sum repayable, *i.e.* K, and (2) the value of the interest—jC—meanwhile receivable, *i.e.* $\frac{j}{i}(C - K)$.

Therefore,

$$A = K + \frac{j}{i}(C - K).$$

If, instead of considering the value of the loan as a whole, it be required to find the value of 1 payable according to the stipulated conditions (which is the form in which the calculation is usually made) then in the above expression C is equal to 1, and K represents the present value of 1, and the formula becomes

$$A = K + \frac{j}{i}(1 - K)$$

$$= 1 - (1 - K) + \frac{j}{i}(1 - K)$$

$$= 1 - (1 - K)\left(1 - \frac{j}{i}\right).$$

F

These general expressions assume different forms according to the manner in which the loan is to be liquidated.

(1) If the loan be repayable in one sum at the end of n years, the value of each unit by the formula we have just found is

$$A = 1 - (1 - v^n)\left(1 - \frac{j}{i}\right)$$

$$= 1 - \frac{1 - v^n}{i}(i - j)$$

$$= 1 - a_{\overline{n}|}(i - j),$$

where $a_{\overline{n}|}$ is taken at rate i.

That this result is correct is easily seen. If the loan were yielding interest at the rate i, which it is desired to realise, the value of each unit would, of course, be 1 ; but as it is only yielding interest at the rate j, it is necessary to deduct from 1 the value of an annuity of the difference of interest.

(2) If, instead of being repaid in one sum at the end of the period, the loan be repayable by n equal instalments of $\frac{1}{n}$ each, K in this case becomes $\frac{a_{\overline{n}|}}{n}$, and the formula for the value of each unit is

$$1 - \left(1 - \frac{a_{\overline{n}|}}{n}\right)\left(1 - \frac{j}{i}\right),$$

where $a_{\overline{n}|}$ is calculated at rate i.

In making use of these formulae the greatest care must be taken that the proper values are attached to the various symbols.

Practical Examples.

(1) Required the annuity which will repay a loan of £1000, in 8 years with 5 per cent. interest.

The formula here is $\dfrac{P}{a_{\overline{n}|}}$, or, substituting the given figures,

$$\frac{1000}{a_{\overline{8}|}} = \frac{1000}{6 \cdot 4632} = 154 \cdot 722.$$

The annuity required is, therefore, £154 14s. 6d.

(2) Required the Sinking Fund to be annually invested at 5 per cent. interest to repay a loan of £1000 on the expiry of 8 years.

For the amount of the sinking fund we found two expressions—

$$P\left(\frac{1}{a_{\overline{n}|}} - i\right), \text{ and } \frac{P}{s_{\overline{n}|}}.$$

Taking the first of these, and substituting the given figures, we have

$$1000\left(\frac{1}{6\cdot463} - \cdot05\right) = £104 \ 14s. \ 6d.$$

By the second formula,

$$\frac{1000}{9\cdot549} = £104 \ 14s. \ 6d.$$

(3) The following schedule shows the working out of the annuity required to repay a loan of £1000 in 8 years with 5 per cent. interest, and the amount of principal and interest respectively contained in each annual payment of £154·722.

Year.	Capital Invested at beginning of each Year.	Interest thereon.	Capital Repaid in each Instalment.	Capital Outstanding at end of each Year.
1	1000·000	50·000	104·722	895·278
2	895·278	44·764	109·958	785·320
3	785·320	39·266	115·456	669·864
4	669·864	33·494	121·228	548·636
5	548·636	27·432	127·290	421·346
6	421·346	21·068	133·654	287·692
7	287·692	14·385	140·337	147·355
8	147·355	7·367	147·355	---

(4) Required the annuity which will repay a loan of £1000 in 8 years with 5 per cent. interest, the Sinking Fund to be accumulated at 3 per cent.

The formula in this case is

$$P\left\{\frac{1}{a'_{\overline{n}|j}} + (i - j)\right\} \quad (a' \text{ being taken at rate } j).$$

At 3 per cent. $\dfrac{1}{a_{\overline{8}|}} = \cdot14245$;

so that, substituting our figures, we have

$$1000 \, (\cdot14245 + \cdot05 - \cdot03) = 1000 \times \cdot16245.$$

The required annuity is, therefore, £162 9s.

(5) A bond for £1000 bearing interest at $4\frac{1}{2}$ per cent. for 20 years, and repayable at maturity with a bonus of 20 per cent. is for sale. What can a purchaser give for it so as to realise 5 per cent. on his investment ?

We could answer this question by taking the value of the bond as made up of (1) the value of £1200 due 20 years hence, or $1200 \times v^{20}$, and (2) the value of an annuity of £45 for 20 years, or $45 \times a_{\overline{20}|}$.

Let us, however, apply Makeham's formula—

$$A = K + \frac{j}{i} \, (C - K).$$

Here $K = 1200 \times v^{20}$, $C = 1200$,

$$i = \cdot05, \qquad j = \frac{1000}{1200} \times \cdot045 = \cdot0375,$$

and at 5 per cent. $v^{20} = \cdot376889$.

We therefore have

$$A = 1200 \times \cdot376889 + \frac{\cdot0375}{\cdot05} \, (1200 - 1200 \times \cdot376889)$$

$$= 1200 \left\{ \cdot376889 + \frac{\cdot0375}{\cdot05} \, (1 - \cdot376889) \right\}$$

$$= 1200 \times \cdot844222 = £1013 \ 1s. \ 4d.$$

(6) A bond for £1000 repayable in 10 annual instalments, and bearing interest at 3 per cent. is for sale. What can a purchaser give for same to realise 5 per cent. on his investment ?

We here use the formula,

$$A = 1 - \left(1 - \frac{a_{\overline{n}|}}{n} \right) \left(1 - \frac{j}{i} \right) ;$$

and, substituting our figures, we have as the value of the bond

$$1000 \left\{ 1 - \left(1 - \frac{a_{\overline{10}|}}{10} \right) \left(1 - \frac{\cdot03}{\cdot05} \right) \right\}$$

$$= 1000 \{ 1 - (1 - \cdot77217)(1 - \cdot6) \}$$

$$= 1000(1 - \cdot091132) = £908 \ 17s. \ 4d.$$

CHAPTER VI.

ANNUITIES-CERTAIN.

DURATION OF ANNUITY, AND RATE OF INTEREST YIELDED.

Given Amount and Rate of Interest to find Duration of Annuity. For the amount of an annuity of 1 payable yearly we found the general formula

$$s = \frac{(1+i)^n - 1}{i}.$$

If the amount and rate of interest be given, and it be desired to find the number of years which the annuity has to run, we transpose the above equation, thus

$$1 + is = (1 + i)^n,$$

whence, $\log (1 + is) = n \log (1 + i)$,

and, $n = \dfrac{\log (1 + is)}{\log (1 + i)}$.

Given Amount and Duration of Annuity to find Rate of Interest. If the amount and the duration of the annuity be given, and it be desired to find the rate of interest yielded, the equation takes a form in which an exact solution is impossible.

To arrive at an approximation to the value of i, we refer to a table of the amounts of annuities, and opposite the given number of years find the amount which corresponds most closely to the given value of s. Say this is s', the rate of interest at which it is calculated being j, we have to find the difference between this rate j and the required rate i. Let us call this difference (which may be positive or negative) ρ. Then

$$i = j + \rho,$$

and, $s = \dfrac{(1+i)^n - 1}{i} = \dfrac{\{(1+j) + \rho\}^n - 1}{j + \rho}$,

whence, $(j + \rho)s = \{(1+j) + \rho\}^n - 1$.

Expanding $\{(1+j)+\rho\}^n$ by the Binomial Theorem, and neglecting all powers of ρ above the first (which we can do without much loss of accuracy as the series is rapidly convergent), we have

$$(j+\rho)s = (1+j)^n + n(1+j)^{n-1}\rho - 1$$

$$= j\left\{\frac{(1+j)^n - 1}{j}\right\} + n(1+j)^{n-1}\rho$$

$$= js' + n(1+j)^{n-1}\rho,$$

$$\therefore \{s - n(1+j)^{n-1}\}\rho = j(s' - s),$$

whence, $\rho = \dfrac{j(s'-s)}{s - n(1+j)^{n-1}},$

and, $i = j + \dfrac{j(s'-s)}{s - n(1+j)^{n-1}}.$

Closer approximations can be found by taking into account further terms in the expansion of $\{(1+j)+\rho\}^n$, but for nearly all practical purposes the above result is sufficiently accurate.

Given the Present Value and Rate of Interest, to find Duration of Annuity. If the present value and rate of interest be given, and it be desired to find the duration of an annuity of 1 payable yearly, we have

$$a = \frac{1 - (1+i)^{-n}}{i},$$

$$\therefore (1+i)^{-n} = 1 - ia,$$

and, $n \log(1+i) = -\log(1-ia),$

whence, $n = \dfrac{-\log(1-ia)}{\log(1+i)}.$

Given the Present Value and Duration of Annuity, to find the Rate of Interest. If now the present value and duration of the annuity be given, and it be required to find the rate of interest, the equation in this case also assumes a form in which a direct solution is impossible, and we can only approximate to the value of i.

As before we refer to the tables, and opposite the given number of years find the annuity value nearest to the given value. Let us call this value as found in the tables a', and the rate of interest at which it is calculated j. We have again to find the difference between this rate j and the required rate i, which difference (positive or negative) we denote by ρ.

We have
$$a = \frac{1 - (1+i)^{-n}}{i}$$

$$= \frac{1 - \{(1+j) + \rho\}^{-n}}{j + \rho},$$

whence, $(j+\rho)a = 1 - \{(1+j) + \rho\}^{-n}.$

We might now proceed, as in the former case when the amount of the annuity was given, to expand $\{(1+j) + \rho\}^{-n}$ by the Binomial Theorem, but a more accurate result will be obtained by first transposing the equation and writing it in the form

$$a = \left(1 - \{(1+j) + \rho\}^{-n}\right) \times (j+\rho)^{-1}.$$

We now expand $\{(1+j) + \rho\}^{-n}$ and $(j+\rho)^{-1}$, and we have

$$a = \left\{1 - (1+j)^{-n} + n(1+j)^{-\overline{n+1}}\rho - \frac{n(n+1)}{2}(1+j)^{-\overline{n+2}}\rho^2 + \text{etc.} \right\}$$
$$\times (j^{-1} - j^{-2}\rho + j^{-3}\rho^2 - \text{etc.}),$$

but $(1+j)^{-n} = v^n$,

therefore,

$$a = \left\{1 - v^n + nv^{n+1}\rho - \frac{n(n+1)}{2}v^{n+2}\rho^2 + \text{etc.} \right\} \times \left(\frac{1}{j} - \frac{\rho}{j^2} + \frac{\rho^2}{j^3} - \text{etc.} \right)$$

$$= \frac{1 - v^n}{j} + \frac{nv^{n+1}}{j}\rho - \frac{1 - v^n}{j^2}\rho - \frac{n(n+1)}{2}\frac{v^{n+2}}{j}\rho^2 - \frac{nv^{n+1}}{j^2}\rho^2 + \frac{1 - v^n}{j^3}\rho^2 - \text{etc.}$$

To get a first approximation to the value of ρ let us neglect the terms involving ρ^2, and we have

$$a = \frac{1 - v^n}{j} + \frac{nv^{n+1}}{j}\rho - \frac{1 - v^n}{j^2}\rho$$

$$= a' + \frac{nv^{n+1}}{j}\rho - \frac{a'}{j}\rho$$

whence, $\rho = \dfrac{j(a' - a)}{a' - nv^{n+1}}$

and, $\boxed{i = j + \dfrac{j(a' - a)}{a' - nv^{n+1}}.}$

This is known as Barrett's Formula, and for nearly all practical purposes it is a sufficiently close approximation. By carrying the work a step further, however, a formula is obtained which gives very accurate results. To do this the value of ρ already found is made use of, and for ρ^2 in the above expansion we substitute

$$\rho \times \frac{j(a'-a)}{a'-nv^{n+1}}.$$

We thus obtain the equation

$$a = \frac{1-v^n}{j} + \frac{nv^{n+1}}{j}\rho - \frac{1-v^n}{j^2}\rho - \frac{n(n+1)}{2}\frac{v^{n+2}}{j} \times \frac{j(a'-a)}{a'-nv^{n+1}}\rho$$

$$- \frac{nv^{n+1}}{j^2} \times \frac{j(a'-a)}{a'-nv^{n+1}}\rho + \frac{1-v^n}{j^3} \times \frac{j(a'-a)}{a'-nv^{n+1}}\rho$$

$$= a' + \frac{nv^{n+1}}{j}\rho - \frac{a'}{j}\rho - \frac{n(n+1)}{2}\frac{v^{n+2}}{j} \times \frac{j(a'-a)}{a'-nv^{n+1}}\rho$$

$$- \frac{nv^{n+1}}{j^2} \times \frac{j(a'-a)}{a'-nv^{n+1}}\rho + \frac{a'}{j^2} \times \frac{j(a'-a)}{a'-nv^{n+1}}\rho.$$

It is unnecessary to show the working out of this equation in detail. From it we obtain

$$\rho = \frac{j(a'-a)}{a - nv^{n+1} + \dfrac{n(n+1)}{2}\dfrac{j(a'-a)v^{n+2}}{a'-nv^{n+1}}},$$

and, $i = j + \rho$.

This formula was originated by J. J. M'Lauchlan, of Edinburgh.

The approximation first obtained, viz.,

$$i = j + \frac{j(a'-a)}{a'-nv^{n+1}},$$

while not so accurate as that just found, is simpler and more easily applied and may in most cases be adopted.

Indeed, without going through this calculation at all, it is sometimes sufficient to ascertain the nearest corresponding annuity value from the tables, and, assuming that the difference in values is proportional to the difference in the rate of interest, find an approximation to the required rate by simple proportion.

A very common application of the foregoing formulae occurs in connection with Foreign Government Loans. These very frequently are issued at a discount, but are redeemed at par by means of an accumulative sinking fund set aside annually for this purpose—bonds to the amount of the sinking fund being drawn and cancelled. Of course bonds drawn within the first year or two will show a much greater

Application of Formulae to Foreign Government Loans.

return than those which are not repaid till the loan has almost expired, and it is impossible to say in the case of any individual bond what the return may be. The rate of interest which the loan as a whole will return to the investors can, however, easily be calculated as follows :—

Let p = the price of issue per unit,

$\qquad$ j = the rate of interest per unit of its par value which the loan yields,

$\qquad$ f = the sinking fund per unit of par value which is annually provided, and which, along with the interest saved on the bonds previously cancelled, is employed in paying off the bonds as they are drawn,

$\qquad$ i = the rate of interest realised by the investors in the aggregate.

The first step is to ascertain the duration of the loan, *i.e.* the number of years that will elapse before all the bonds are drawn and cancelled. Now it is clear that to employ the sinking fund together with the interest saved in paying off bonds, is the same in effect as to accumulate the sinking fund at the nominal rate of interest which the loan yields. To find the duration of the loan we accordingly have the equation

$$1 = f \cdot \frac{(1+j)^n - 1}{j},$$

whence, $\quad n = \dfrac{\log\left(\dfrac{j}{f} + 1\right)}{\log\,(1+j)}.$

Having thus found the duration of the loan, we can obviously regard p (the price of issue) as the value of an annuity of $(j+f)$ for n years,

whence, $\quad p = (j+f)\,a_{\overline{n}|},$

or, $\quad a_{\overline{n}|} = \dfrac{p}{j+f}.$

The present value of an annuity of 1 and the number of years it has to run being thus ascertained, i can be found by applying one of the formulae of approximation.

Determination of Rate of Interest by Makeham's Formula. For finding the rate of interest which loans with more or less complicated conditions of repayment yield to investors Makeham's formula, which was investigated in last chapter, is available.

For the value of a loan we found the general expression

$$A = K + \frac{j}{i}(C - K),$$

whence, $i(A - K) = j(C - K)$,

and, $i = j\dfrac{C - K}{A - K}.$

In using this formula, however, to find the value of i we are met at the outset by a practical difficulty. K (the present value of C) is calculated at the rate i and cannot be ascertained till i is known. This is overcome by assuming a value for i as nearly correct as possible, and calculating K at this assumed rate. The value of K thus obtained is then inserted in the formula, and if the resulting value of i closely correspond with the assumed rate, we know that our assumption is correct. If it should not correspond with sufficient exactness, the process is repeated, and a close approximation to the value of i is obtained by interpolation.

Practical Examples.

(1) In how many years will an annuity of £40 amount to £4832 at 3 per cent. interest ?

The formula here is $n = \dfrac{\log (1 + is)}{\log (1 + i)}$,

and, as $s = \dfrac{4832}{40}$, and $i = \cdot03$,

we have $n = \dfrac{\log \left(1 + \dfrac{4832}{40} \times \cdot03\right)}{\log 1\cdot03} = \dfrac{\log 4\cdot624}{\log 1\cdot03}$,

but, $\log 4\cdot624 = 0\cdot6650178$, and $\log 1\cdot03 = 0\cdot0128372$,

$$\therefore\ n = \frac{\cdot6650178}{\cdot0128372} = 51\cdot8 \text{ years.}$$

(2) At what rate per cent. will £20 per annum amount in 10 years to £243 ?

In this case $$s = \frac{243}{20} = 12\cdot15,$$

and referring to the tables we find that in 10 years at 4 per cent.
the amount of an annuity of 1 is 12·006.

Accordingly, $s' = 12·006$, and $j = ·04$.

We now apply the formula

$$i = j + \frac{j(s' - s)}{s - n(1+j)^{n-1}} \quad = ·04 + \frac{·04(12·006 - 12·15)}{12·15 - 10 \times (1·04)^9}$$

$$= ·04 + \frac{- ·00576}{12·15 - 14·233} = ·04 + \frac{·00576}{2·083}$$

$$= ·04277, \text{ or, } 4·277 \text{ per cent.}$$

(3) For how many years can an annuity of £30 be purchased for £551 15s., interest being taken at $3\frac{1}{2}$ per cent. ?

The formula here is $n = \dfrac{-\log(1 - ia)}{\log(1 + i)}$,

and, as $a = \dfrac{551·75}{30}$, and $i = ·035$,

we have $n = \dfrac{-\log\left(1 - ·035 \times \dfrac{551·75}{30}\right)}{\log 1·035}$

$$= \frac{-\log ·35629}{\log 1·035},$$

but, $\log ·35629 = \bar{1}·5518036$, $\therefore -\log ·35629 = 0·4481964$,

and, $\log 1·035 = 0·0149403$,

$$\therefore n = \frac{·4481964}{·0149403} = 30 \text{ years.}$$

(4) An annuity having 30 years to run is sold at 19 years' purchase; what rate of interest does it yield to the purchaser?

In this case $\qquad a = 19$, and $n = 30$,

and turning to the tables, we find that the present value of an
annuity of 1 for 30 years at $3\frac{1}{2}$ per cent. interest is 18·392;

accordingly, $\qquad a' = 18·392$, and $j = ·035$.

We now apply the first formula of approximation, which is
sufficiently accurate

$$i = j + \frac{j(a' - a)}{a' - nv^{n+1}} \quad = ·035 + \frac{·035(18·392 - 19)}{18·392 - 30(1·035)^{-31}}$$

$$= ·035 + \frac{- ·02128}{18·392 - 10·327} = ·035 - ·00264$$

$$= ·03236, \text{ or } 3·236 \text{ per cent.}$$

(5) A Foreign Government borrows £2,000,000 at 5 per cent. interest with an Accumulative Sinking Fund of 2 per cent. for the redemption of the bonds at par. The issue price is 90 per cent. What is the actual rate of interest paid by the borrowers ?

We have first to find the number of years that will elapse till the loan is paid off.   To do this we apply the formula

$$n = \frac{\log\left(\dfrac{j}{f} + 1\right)}{\log\left(1 + j\right)}.$$

Now $j = ·05$, and $f = ·02$, and we have accordingly

$$n = \frac{\log\left(\dfrac{·05}{·02} + 1\right)}{\log 1·05} = \frac{0·5440680}{0·0211893}$$

$$= 26 \text{ years nearly.}$$

Note.—It is usually sufficient to take the nearest integral value for n.

The present value of each unit of the yearly payments to be made on account of the loan is

$$a = \frac{p}{j+f} = \frac{·9}{·07} = 12·857.$$

We are now in a position to apply one of the approximate formulae for the rate of interest ; and, turning to the tables, we find that the present value of an annuity of 1 for 26 years at 6 per cent. is 13·003.

By Barrett's formula

$$i = j + \frac{j(a' - a)}{a' - nv^{n+1}},$$

$$= ·06 + \frac{·06(13·003 - 12·857)}{13·003 - 26(1·06)^{-27}}$$

$$= ·06 + \frac{·00876}{13·003 - 5·392} = ·06 + ·00115.$$

The rate of interest is thus 6·115 per cent.

Let us now take M'Lauchlan's formula, which, although more troublesome to work with, should give a more accurate result.

The formula is

$$i = j + \frac{j(a' - a)}{a - nv^{n+1} + \dfrac{n(n+1)}{2} \dfrac{j(a' - a)v^{n+2}}{a' - nv^{n+1}}}.$$

To obtain the value of the term

$$\frac{n(n+1)}{2} \frac{j(a' - a)v^{n+2}}{a' - nv^{n+1}},$$

we have already found that

$$\frac{j(a' - a)}{a' - nv^{n+1}} = \cdot00115,$$

and we have to multiply this by $\dfrac{26 \times 27}{2}$, and by $(1\cdot06)^{-28}$.

We thus get as the value of this expression $\cdot079$.
Accordingly

$$i = \cdot06 + \frac{\cdot00876}{12\cdot857 - 5\cdot392 + \cdot079},$$

$$= \cdot06 + \cdot00116.$$

This result is almost identical with that already found; but if there had been a greater difference between the values of a and a', the results obtained by the two formulae would not have agreed so closely.

(6) What proportion of the above loan remains outstanding after the seventh payment has been made, and if the loan be then quoted at 97 per cent., what rate of interest does it yield?

The annual sum payable under the conditions of the loan being £7 for every £100 borrowed, the capital outstanding after the seventh payment is $7 \times a_{\overline{19}|}$, or £84·597, for every £100 originally borrowed (the annuity being calculated at 5 per cent. interest). As the loan is then quoted at 97 per cent., the market value of an annuity of 1 for 19 years will be

$$a_{\overline{19}|} = \frac{84\cdot597}{7} \times \frac{97}{100},$$

$$= 11\cdot722;$$

and applying the first formula of approximation, we find that this represents a rate of interest of about 5·37 per cent.

(7) What rate of interest would the Government Loan we have been considering yield if, instead of being redeemable by an Accumulative Sinking Fund, it were to be redeemed at par by uniform annual drawings of 2 per cent. ?

To answer this question, let us take Makeham's formula and assume as a trial rate 6 per cent. interest. We have

$$i = j \frac{C - K}{A - K}$$

$$= \cdot 05 \times \frac{100 - 2a_{\overline{50}|}}{90 - 2a_{\overline{50}|}} \quad \text{(calculating } a_{\overline{50}|} \text{ at 6 per cent.)}$$

$$= \cdot 05 \times \frac{100 - 31 \cdot 524}{90 - 31 \cdot 524}$$

$$= \cdot 0585.$$

It thus appears that the trial rate is too high. Repeating the calculation at $5\frac{3}{4}$ per cent., we have

$$i = \cdot 05 \times \frac{100 - 2a_{\overline{50}|}}{90 - 2a_{\overline{50}|}} \quad \text{(calculating } a_{\overline{50}|} \text{ at } 5\frac{3}{4} \text{ per cent.)}$$

$$= \cdot 05 \times \frac{100 - 32 \cdot 655}{90 - 32 \cdot 655}$$

$$= \cdot 0587.$$

From these assumed rates a close approximation to the true rate is found by interpolation as follows :—

$$5 \cdot 75 \text{ gives } 5 \cdot 87$$
$$6 \cdot 00 \quad , \quad 5 \cdot 85$$

Differences $\quad \cdot 25 \qquad - \cdot 02$

whence, $5 \cdot 75 + x = 5 \cdot 87 - \dfrac{\cdot 02}{\cdot 25} x$ very nearly,

and, $x = \dfrac{5 \cdot 87 - 5 \cdot 75}{1 + \dfrac{\cdot 02}{\cdot 25}} = \dfrac{\cdot 12}{1 \cdot 08} = \cdot 111.$

Accordingly, the approximate rate of interest yielded by the loan is $5 \cdot 75 + x$, or $5 \cdot 86$ per cent.

PART II.

LIFE CONTINGENCIES.

PART II.

LIFE CONTINGENCIES.

CHAPTER I.

THE MORTALITY TABLE.

METHODS OF CONSTRUCTION AND ADJUSTMENT.
PRINCIPAL TABLES IN USE.

Construction of Mortality Tables. UNCERTAIN as is the lifetime of any individual, it is found that the average duration of life in a large community is subject to little fluctuation. If, therefore, a table of mortality be formed from the collected experience of a large number of persons, the results obtained will *on the average* be correct and reliable.

l_x and d_x Columns. In a *Mortality Table*, as ordinarily constructed, there are two elementary columns from which the others are derived, and on which all calculations are based. The first of these, which is known as the l_x column (the suffix x being used to denote the age) begins with an arbitrary number of persons supposed to be alive, and shows how many out of this number will survive at each year of age till the limit of life is reached. The second, known as the d_x column, gives the number who die at each age. The arbitrary number with which the l_x column starts is called the *radix* of the table.

While we have treated these two columns as distinct, they in reality present the same facts from different points of view. If there be 100 persons alive at age 10, and at age 11 only 97 of these survive, the number who have died aged 10 is of course $100 - 97$.

We accordingly have the relation

$$l_{10} - l_{11} = d_{10},$$

or, to put the case generally

$$l_x - l_{x+1} = d_x.$$

Again, it is obvious that if the d_x column be added from the last age upwards the sum will be equal to l_x, and we have therefore the further relation that the number in the l_x column at any age is equal to the sum of the d_x column from that age to the end of the table.

Number Dying between Ages x and $x+n$. To find the number dying between ages x and $x+n$ (1) the values in the d_x column, from d_x to d_{x+n-1} inclusive, may be added, or (2) the same result may be obtained more simply by subtracting the number alive at age $(x+n)$ from l_x. That is

$$d_x + d_{x+1} + d_{x+2} + \ldots + d_{x+n-1} = l_x - l_{x+n}.$$

Limiting Age and Complement of Life. The values in the l_x column, representing the numbers who out of the given *radix* attain each year of life, will, of course, decrease year by year as age advances, until an age is reached which some enter upon but none complete. This is called the *limiting age* of the table, and the difference between the limiting age and the present age is known as the *complement of life*. To denote the *limiting age* the suffix ω is used, and accordingly $l_\omega = 0$. The *complement of life* as defined above is $(\omega - x)$.

In addition to these elementary columns, the Mortality or Life Table (as it is sometimes called) usually contains others showing the probabilities of living and dying and the expectations of life. We shall see in a subsequent chapter how these are deduced, but meanwhile we have to consider how the fundamental columns of the mortality table are constructed.

Two methods of forming these will at once suggest themselves—

(1) They may be based on actual facts.

(2) An attempt may be made to formulate a law of life and found the table upon it.

Mortality Table deduced from an assumed Law of Life. The consideration of the latter of these methods involves questions of much interest, but is of comparatively little practical importance. Of late years, however, great advances have been made in the development of a theory of life, and the most recent tables (while founded on facts) are graduated in this way.

Before proceeding, therefore, to consider how a table may be deduced from actual experience, we shall refer shortly to the attempts that have been made to establish a law of life.

De Moivre's Hypothesis.

The first important step in this direction was taken by De Moivre, who in 1725 published his famous hypothesis of equal decrements of human life. Starting with a group of persons of age 12, he assumed that the same number would die each year, and that by age 86 (which he took as the limit of life) all would be dead. According to this hypothesis the number living at each age is in decreasing arithmetical progression. Thus assuming 74 to be alive at age 12, 73 would survive at age 13, 72 at age 14, and so on. At age 85 there would be only one survivor, and at 86 he too would have died.

This hypothesis of De Moivre does not (and did not at the time it was originated) represent even approximately the law of human life. For many years, however, it was very extensively used, as it afforded great facilities for making calculations. It is now completely superseded and is of interest only as a matter of history.

Gompertz's Theory of Life.

The next attempt to formulate the law of human life was made by Benjamin Gompertz, who read a paper describing his views before the Royal Society in 1825. His idea was that liability to death is attributable to two co-existing causes—(1) chance without previous disposition to death (the young man being liable to be killed by accident or acute disease as well as the old), and (2) a gradual deterioration, by which, as a man advances in years, he becomes more and more unable to withstand destructive forces.

Upon this hypothesis Gompertz based certain formulae, but they did not fully bear out his theory, and are of little practical value. In recent years, however, the idea has been further developed by the late W. M. Makeham, and formulae for practical working have been deduced which agree very closely with the actual facts of human life as established by recent investigations.

These formulae belong to the higher branches of actuarial science, and cannot here be investigated. It may be mentioned, however, that the formula Makeham deduces for the function l_x is $ks^x g^{c^x}$, where k, s, g, and c are all constants, and x—the age—is the only variable. When these constants have been adjusted to a particular table, it is only necessary to substitute the given age for x in the above expression to find the number living at that age. The factor g^{c^x} involving, as it does, a power of a power, is very peculiar, and at first sight would be thought too complicated to be of practical utility. The double power is easily got rid of, however, by using logarithms.

The main advantage of employing this method in forming and graduating mortality tables is that great facilities are thereby afforded for the calculation of joint life annuities.

We pass now to consider the construction of a mortality table founded on actual observations. Tables are formed in this way to show the mortality experienced either (1) among a particular class—such, for example, as assured lives—or, (2) throughout the population generally.

Construction of Table from actual observations.

If the intention of the table be to show the rate of mortality prevailing among a special class, full particulars are collected regarding a large number of individuals similarly circumstanced. By extracting these particulars on cards, great facilities are afforded for arranging and rearranging the statistics as may be required.

The numbers exposed to risk at each age, and the deaths which have happened amongst them, are thus ascertained, and by dividing the deaths at each age by the number exposed to risk the rate of mortality for that age is arrived at.

When the rate of mortality at each age has in this way been obtained, the l_x and d_x columns may be formed as follows:—

If it be desired to begin the table say at age 10, the l_x column is started with a purely arbitrary number. Multiplying this number by the rate of mortality at age 10 we obtain d_{10}, the initial value of the d_x column. Subtracting d_{10} from l_{10} we have l_{11}, and multiplying this in turn by the rate of mortality at age 11 we get d_{11}. We then subtract d_{11} from l_{11}, which gives l_{12}, and proceeding in this way the table is formed.

If the table is to be based on the mortality experience of the whole community, we proceed somewhat differently.

From published statistics, such, for example, as census returns, the numbers living at each age are ascertained, and the deaths that occur, say for the six months preceding and the six months following the date of the return, are taken from the registers.

By dividing the number of deaths at each age by the number living, a function is obtained from which the rate of mortality is easily deduced, and l_x and d_x columns are formed in the manner explained above. Care must be taken that the statistics on which the table is based have not been unduly affected by an epidemic or other exceptional cause.

Graduation.

A table constructed in either of these ways will, of course, show very varying results, and will be of no practical use till it has been graduated, so that the figures from age to age may run in regular progression.

There is great difference of opinion amongst actuaries as to the best means of adjusting tables, and some of the methods which have been devised, while giving a smooth progression, are apt to destroy outstanding characteristics at particular ages which ought to be retained. The simplest plan is that adopted by Mr. Finlaison, the Actuary to the National Debt Commissioners, which consists in summing the ungraduated figures in groups of five, taking the average, and placing the figure so obtained opposite the central age of the group. This process is repeated till a satisfactory adjustment is obtained.

Error in assuming equal distribution of deaths.

In deducing the probabilities of life from a table formed in the manner we have described an element of error (though not of serious error) is introduced into the calculations.

The results obtained would only be strictly accurate on the assumption that the deaths were equally distributed over each year —an assumption which is manifestly not in accordance with facts.

Formulae have been devised by which, with the aid of the Differential and Integral Calculus, this assumption is entirely got rid of, but the error involved is of no practical importance.

Northampton Table.

We shall now consider briefly the principal Life Tables in use, and in so doing we must first refer to *The Northampton Table* constructed by Dr. Price in 1783. This was practically the first, and for many years the only life table, but being based on a false assumption it was found to be grossly misleading, and it is now almost entirely disused. Dr. Price assumed that the population of All Saints Parish, Northampton (from the statistics of which he compiled his table) was stationary, and that an apparent excess in the number of deaths over the number of births was due to immigration. This assumption was erroneous, the real cause of the discrepancy being that, on account of the large number of Baptists in the parish, the Register of Baptisms did not show the true numbers born. For this reason the table exhibits a very excessive rate of mortality at the younger ages.

Carlisle Table.

The Carlisle Table was constructed by Mr. Milne, Actuary of the Sun Life Office, and published in 1815. This table was carefully and correctly formed, and is still very extensively used for calculating life annuities. About 55 per cent. of those under observation were females, and the table accordingly represents in a marked degree the leading characteristics of female life, chief among these being great vitality at the older ages.

Government Life Annuitants' Tables.

We next notice *The Government Life Annuitants' Tables.* In 1808 the National Debt Commissioners began to grant life annuities, and special tables were prepared for them based on the Northampton Table. As we have seen, however, the Northampton Table greatly exaggerates the rate of mortality at the younger ages, and in 1819 the Government became alive to the fact that they were losing heavily. Indeed it was pointed out to them that in 11 years they had lost nearly £2,000,000. As a result the Government instructed Mr. Finlaison to prepare a table showing the true rate of mortality amongst annuitants.

This table was published in 1829, and in 1860 a second was formed on similar lines, and again in 1883 a third, which is the table now in use. The 1883 table was based entirely on the records of the National Debt Office from 1808 to 1875. These tables show in a marked degree the superiority of female life at the older ages.

For the calculation and valuation of annuities, the 1883 Government Table is considered the most reliable, although the Carlisle Table is still also much used.

We now come to the tables upon which the Life Insurance Offices base their calculations.

Seventeen Offices' Experience Table.

The first table constructed for this purpose was that known as the *Seventeen Offices' Experience*, published in 1843.

The principal object of the investigation, which resulted in the formation of this table, was to ascertain how far the rate of mortality is affected by the medical examination to which those insured are subjected.

It was found that at all ages the mortality among assured females is greater than among males, and also that the medical examination loses its beneficial effect in five years, after which period the rate of mortality among assured lives, instead of being less than that prevailing throughout the population in general, is if anything rather greater. This last result is very singular, but the fact seems well established.

Institute of
Actuaries'
Life Tables.

In 1862 the Seventeen Offices' Experience Tables were. superseded by what are known as the *Institute of Actuaries' Life Tables*, and these are now recognised as the standard tables upon which Life Insurance Offices, for the most part, base their premiums and make their valuations.

The tables known under this name are three in number, viz.—

(1) The H^M Table, founded on healthy male lives,

(2) The H^F Table, founded on healthy female lives, and

(3) The $H^{M(5)}$ Table, from which the first five years of insurance are excluded, so as to annul the effect of the medical examination.

It may be noted that the rate of mortality in the H^F Table under age 45 is greatly in excess of that in the H^M Table.

Of late years very exhaustive tables have been compiled in America, founded on the experience of the offices there, but these are not much (if at all) used in this country.

English Life
Tables.

There is another set of tables—the English Life Tables—to which reference must be made. These were compiled by Dr. Farr from the census returns and the register of deaths, and accordingly represent the rate of mortality among the population at large. There are three of these tables, but the *English Life Table No. 3*, published in 1864, is recognised as the standard.

Quite recently a fourth investigation was made, but the results agreed so closely with those previously obtained that it was considered unnecessary to compile fresh tables from them.

CHAPTER II.

THE MORTALITY TABLE.

ITS USE IN DEALING WITH STATISTICAL QUESTIONS.

Population of Stationary Community.

IN dealing with a community in a perfectly stationary condition, undisturbed by emigration or immigration, and with the birth rate and the death rate exactly equal, it might be thought that the l_x column of the Mortality Table could be formed by simply taking a census of the population. Thus, if such a community were recruited by 100,000 births each year, for l_0 there would be the 100,000 just born, for l_1 the survivors of the 100,000 born a year ago, for l_2 the survivors of those born two years ago, and so on, the *total population* being the sum of the l_x column.

This result, however, would only be correct on the assumption that all the births occurred simultaneously, whereas it is evidently much more in accordance with fact to suppose that they are spread over the year.

Now l_x represents the number alive at the *exact* age x, but as has just been stated, the population at age x in such a community would not be of the exact age x, but of every fractional age between x and $x+1$. The population at age x is not therefore represented by l_x, but by $\dfrac{l_x + l_{x+1}}{2}$,—the mean between l_x and l_{x+1}.

The *Total Population* of the stationary community supported by l_0 annual births is accordingly

$$\tfrac{1}{2}(l_0 + l_1) + \tfrac{1}{2}(l_1 + l_2) + \tfrac{1}{2}(l_2 + l_3) + \text{ etc.}$$
$$= \tfrac{1}{2}l_0 + l_1 + l_2 + l_3 + \text{ etc.},$$

or, using the symbol L_x for the population aged x, so that

$$L_x = \tfrac{1}{2}(l_x + l_{x+1}),$$

the *Total Population* in this notation is

$$L_0 + L_1 + L_2 + L_3 + \text{ etc.}$$

If a column of these values be formed, and if this be summed from the limiting age of the table upwards, the results being placed in another column T_x, so that $T_x = L_x + L_{x+1} + L_{x+2} +$ etc., it is possible with the aid of these values to deal by simple proportion with many statistical questions relating to population.

Thus, for example, if it be required to determine the number of inhabitants in a village in which 100 births take place annually, and which is undisturbed by emigration, since the total population which l_0 annual births will support is T_0, the population supported by 100 annual births, or in other words the number of inhabitants in the village, will be

$$\frac{100}{l_0} \times T_0.$$

Similarly, if a society be recruited annually by 100 entrants of age 20, the total membership of the society when it has reached a stationary condition will be

$$\frac{100}{l_{20}} \times T_{20}.$$

And if the members of this society have to resign at 60 years of age, the total number on the roll at any time will be

$$\frac{100}{l_{20}} (T_{20} - T_{60}).$$

As the community under consideration is supposed

Total Annual Deaths in Stationary Community.

to be in a stationary condition, the number who die annually must be equal to the number born. The number of annual deaths throughout the whole population—T_0—is accordingly l_0, and the death rate per cent. is

$$\frac{100 \times l_0}{T_0}.$$

To find the average age at death in the community:—

Average Age at Death.

Of the l_0 children born, l_1 will survive a year, and those who die will on the average have lived for half a year. Accordingly, the total number of years lived by the l_0 children in the first year will be

$$l_1 + \tfrac{1}{2} d_0 = l_1 + \frac{l_0 - l_1}{2} = \frac{l_0 + l_1}{2} = L_0.$$

Similarly, in the second year of their life the aggregate existence of the l_0 persons will be L_1, and in the third year L_2. The total number of years lived by the l_0 persons will therefore be

$$L_0 + L_1 + L_2 + \text{etc.} = T_0,$$

and accordingly their average age at death will be

$$\frac{T_0}{l_0}.$$

To find the average age at death of the population at present aged x :—

If we follow the same method of procedure, it is evident that the average future existence of the persons aged x will be

$$\frac{T_x}{l_x},$$

and, adding to this the x years already lived by them, their average age at death will be

$$x + \frac{T_x}{l_x}.$$

From the above it is apparent that the higher the age on entry the greater will be the average age at death, and that no conclusions can be drawn from the average ages at death of different groups of persons, unless due allowance be made for the different ages at which they come under observation.

Results only applicable to Society in Stationary Condition.

The foregoing results are only applicable to a community in a *stationary condition*, *i.e.* one in which there is a due proportion of members at all ages, and in order that this description may apply, it is necessary not only that the community should have been in existence for many years undisturbed by emigration or immigration, but also that the number born into the community, or annually admitted at the age of entry, should always be the same.

In point of fact, hardly any society or community will fulfil these requirements, and modifications have to be made on the formulae to suit the special circumstances of each case.

By ignoring the preliminary condition that the community under consideration must be stationary if the principles above investigated are to apply, erroneous conclusions are sometimes deduced.

It is often said, for instance, that one district of the country is more healthy than another because the death rate per 1000 is lower. If, however, there be for any reason an influx of young lives into one of these districts, or if there be a disturbance by emigration of the relative proportions of the population at different ages it is obvious that no such conclusion can fairly be drawn.

It has been argued also, that because the average annual death rate of an old established insurance society is say 10 per 1000 of those insured, while the average premium charged is over £20 per £1000, the company is charging twice as much as is necessary to cover its risks. In arguing in this way no account is taken of the fact that in recent years the society has probably doubled or trebled its business, and has accordingly a large preponderance of young lives on its books, by which the average death rate among its members is necessarily much affected.

An example illustrating the practical construction of the L_x and T_x columns will be found at the end of the next chapter.

CHAPTER III.

LIFE CONTINGENCIES.

PROBABILITIES OF LIVING AND DYING.
EXPECTATIONS OF LIFE.

IN a former chapter we saw how the l_x and d_x columns of the mortality table are constructed. Starting with an arbitrary number (called the *radix* of the table) the l_x column shows how many attain each year of life, while the d_x column gives the number who die at each age.

We have now to consider how the probabilities of living and dying and the expectations of life are derived from these elementary values.

The probability of a person aged x living one year is usually represented by the symbol p_x; of his living two years by $_2p_x$, and of his living n years by $_np_x$. Care must be taken not to confound this last symbol with p_{x+n}, which denotes the probability of a person aged $x+n$ living one year.

Probability of (x) living one year. Let us first see how the probability of a person aged x living one year may be determined. This probability is represented in the Life Table by the column p_x.

If it be known that, out of 1000 persons of the same age alive at the beginning of a year, 950 will live to the end of it, the probability that any one of these 1000 persons will live a year is $\frac{950}{1000}$, or (to state the case generally) if, out of l_x persons alive at age x, l_{x+1} attain to age $x+1$, the probability that any one of these l_x persons will live a year is manifestly $\frac{l_{x+1}}{l_x}$.

This is the principle upon which the p_x column of the Life Table is constructed, and we have accordingly

$$p_x = \frac{l_{x+1}}{l_x},$$

$$p_{x+1} = \frac{l_{x+2}}{l_{x+1}},$$

$$p_{x+2} = \frac{l_{x+3}}{l_{x+2}},$$

and so on to the end of the table.

Probability of (x) living n years. To ascertain the probability of a person aged x living n years, the same process of reasoning is employed. Out of l_x persons alive at age x, l_{x+n} will attain age $x+n$. The probability that any one of the l_x persons will live n years is accordingly $\frac{l_{x+n}}{l_x}$, that is

$$_n p_x = \frac{l_{x+n}}{l_x}.$$

Given the p_x Column to construct the Mortality Table. There is a further relation between the l_x and p_x columns, by which, if the successive values of p_x be known the mortality table can be constructed.

We have

$$l_{x+1} = l_x \times \frac{l_{x+1}}{l_x} \qquad = l_x \times p_x,$$

$$l_{x+2} = l_x \times \frac{l_{x+1}}{l_x} \times \frac{l_{x+2}}{l_{x+1}} = l_x \times p_x \times p_{x+1},$$

and similarly

$$l_{x+n} = l_x \times \frac{l_{x+1}}{l_x} \times \ldots \times \frac{l_{x+n}}{l_{x+n-1}} = l_x \times p_x \times \ldots \times p_{x+n-1}.$$

If we start, therefore, with the initial value of the l_x column (which is arbitrary) and multiply it by the probability of living one year, the second value of l_x is obtained; and, by repeating this process an l_x column is formed corresponding to the given values of p_x.

Probability of (x) dying within a year. We have now to investigate the probability of dying at each age, which is represented in the Life Table by the column q_x. The value of this probability may be obtained in two ways—

(1) Of the l_x persons alive at age x, d_x will die before reaching age $x+1$. The probability of any one of these persons dying is accordingly $\dfrac{d_x}{l_x}$, and we have therefore

$$q_x = \frac{d_x}{l_x}.$$

(2) In questions of probability, certainty is always represented by unity. It is a certainty that a person aged x will either be alive or dead at age $x+1$. For the chance of his being dead we therefore have—

certainty, minus the chance of his being alive.

Accordingly

$$q_x = (1 - p_x).$$

The probability of a person dying within a year (or within any number of years) can thus at once be obtained by subtracting from 1 the probability of his being alive at the end of the period.

If, therefore, the p_x column be given, the values of q_x can very easily be deduced.

Let us now consider how the function known as the **Curtate and Complete Expectations of Life.** *Expectation of Life* is derived.

We have seen that out of l_x persons of the exact age x, l_{x+1} will live for one year; l_{x+2} for a second year; l_{x+3} for a third, and so on to the limit of life. In the aggregate, therefore, the l_x persons will live between them—

$$l_{x+1} + l_{x+2} + l_{x+3} + \text{etc. years,}$$

which gives an average future lifetime for each of—

$$\frac{l_{x+1} + l_{x+2} + l_{x+3} + \text{etc.}}{l_x} \text{ years.}$$

This is called the *Curtate Expectation of Life*, and is denoted by the symbol e_x.

No account is here taken of the time lived by each in the year of death, which on the average may be assumed to be half a year. Adding this to the former expression we have

$$\tfrac{1}{2} + \frac{l_{x+1} + l_{x+2} + l_{x+3} + \text{etc.}}{l_x}.$$

This is known as the Complete Expectation of Life, and the symbol for it is $\overset{o}{e}_x$.

From what has been said it is evident that

$$\overset{o}{e}_x = e_x + \tfrac{1}{2}.$$

Given the Expectations of Life, to construct the Mortality Table. It is sometimes necessary to ascertain the mortality table upon which given expectations of life are based. This can easily be done as follows :—

$$e_x = \frac{l_{x+1} + l_{x+2} + l_{x+3} + \text{etc.}}{l_x}$$

$$= \frac{l_{x+1}}{l_x} \times \frac{l_{x+1} + l_{x+2} + l_{x+3} + \text{etc.}}{l_{x+1}}$$

$$e_x = p_x \times (1 + e_{x+1}),$$

$$\therefore p_x = \frac{e_x}{1 + e_{x+1}}.$$

If, therefore, the expectation of life at any age be divided by the expectation at the following age plus 1, the value of p_x is obtained, and when the p_x column has thus been constructed, the mortality table can easily be formed by the method already explained.

The term *Expectation of Life* is somewhat misleading, as the function to which this name is given has no relation whatever to the probable lifetime of any individual. It is simply the average after-lifetime of all the persons of the same age.

Most Probable After-lifetime. The *most probable after-lifetime* of any individual is the difference between his present age and the age at which most deaths occur, and is therefore very different from the expectation of life. At birth the most probable after-lifetime is less than a year, while the expectation of life is from 40 to 50 years.

Vie Probable. Another function with which the expectation of life must not be confounded is the *Vie Probable*, by which is meant the number of years that a person at a given age has an even chance of living, or, in other words, the period that will elapse until the number surviving at the given age is reduced to one-half.

The functions known as *Vie Probable* and *Expectation of Life* would only coincide if De Moivre's hypothesis (see page 57) were correct, *i.e.* if the number dying at each age were always the same.

There are thus three functions which are perfectly distinct, but which are apt to be confounded.

(1) The *Expectation of Life*, which is the average duration of life among a number of persons of a specified age,

(2) The *Vie Probable*, which is the number of years which a person of a given age has an even chance of living, and

(3) The *Most Probable After-lifetime*, which is the difference between the present age of any individual and the subsequent age at which most deaths occur.

Summary of principal results. The principal relations we have established may now be summarized as follows :—

$$p_x = \frac{l_{x+1}}{l_x} = (1 - q_x),$$

$$q_x = \frac{d_x}{l_x} = (1 - p_x),$$

$$_n p_x = \frac{l_{x+n}}{l_x} = (1 - /_n q_x),$$

$$/_n q_x = \frac{l_x - l_{x+n}}{l_x} = (1 - {_n p_x}),$$

$$e_x = \frac{l_{x+1} + l_{x+2} + l_{x+3} + \text{etc.}}{l_x} = p_x + {_2 p_x} + {_3 p_x} + \text{etc.,}$$

$$\overset{\bullet}{e}_x = \tfrac{1}{2} + \frac{l_{x+1} + l_{x+2} + l_{x+3} + \text{etc.}}{l_x} = \tfrac{1}{2} + e_x.$$

Joint Life Probabilities. Let us now look very briefly at the *Joint Life Probability* that a person aged x and another aged y will both survive one year.

As there are l_x persons aged x, and l_y persons aged y, by pairing each of the former in turn with each of the latter $l_x \times l_y$ pairs are formed. A year hence there will be l_{x+1} persons aged $(x+1)$ and l_{y+1} aged $(y+1)$, and these will in the same way constitute $(l_{x+1} \times l_{y+1})$ pairs. The probability, therefore, of any pair remaining unbroken for a year, or in other words the probability of a person aged x and another aged y both living for a year is

$$\frac{l_{x+1} \times l_{y+1}}{l_x \times l_y} = p_x \times p_y = p_{xy}.$$

Similarly it might be shown that

$$_n p_{xy} = \frac{l_{x+n} \times l_{y+n}}{l_x \times l_y} = \frac{l_{x+n:y+n}}{l_{xy}}.$$

We have likewise

$$|_n q_{xy} = (1 - {_n}p_{xy}),$$

and for the joint expectation of life

$$e_{xy} = \frac{l_{x+1:y+1} + l_{x+2:y+2} + \text{etc.}}{l_{xy}}$$

$$= p_{xy} + {_2}p_{xy} + {_3}p_{xy} + \text{etc.}$$

It must be noted, however, that for the complete expectation of life we cannot write

$$\overset{\circ}{e}_{xy} = \tfrac{1}{2} + e_{xy},$$

as in the case of single lives.

Practical Examples.

(1) In a community 2000 marriages are contracted, the men being 30 years of age and the women 25. Find how many will survive as married couples at the end of 10 years, the probability of dying within 10 years at age 30 being ·106, and at age 25, ·09.

The probability of living for 10 years at age 30 is $(1 - \cdot106)$, and at age 25, $(1 - \cdot09)$.

The number of married couples at the end of the 10 years will therefore be

$$2000 \times (1 - \cdot106) \times (1 - \cdot09)$$

$$= 2000 \times \cdot894 \times \cdot91$$

$$= 1627.$$

(2) *Given the l_x column from age 60 onwards as undernoted, construct columns for d_x, L_x, T_x, p_x, q_x, e_x, and $\overset{\circ}{e}_x$.*

x	l_x	d_x	L_x	T_x	p_x	q_x	e_x	$\overset{\circ}{e}_x$
Age.	Number Living.	$l_x - l_{x+1}$	$\dfrac{l_x + l_{x+1}}{2}$	ΣL_x	$\dfrac{l_{x+1}}{l_x}$	$1 - p_x$	$\dfrac{\Sigma l_{x+1}}{l_x}$	$e_x + \cdot 5$
60	362	12	356	5188	·97	·03	13·8	14·3
61	350	12	344	4832	·96	·04	13·3	13·8
62	338	12	332	4488	·96	·04	12·8	13·3
63	326	12	320	4156	·96	·04	12·2	12·7
64	314	12	308	3836	·96	·04	11·7	12·2
65	302	12	296	3528	·96	·04	11·2	11·7
66	290	12	284	3232	·96	·04	10·6	11·1
67	278	12	272	2948	·96	·05	10·1	10·6
68	266	13	259	2676	·95	·05	9·5	10·0
69	253	13	247	2417	·95	·05	9·0	9·5
70	240	13	233	2170	·95	·05	8·5	9·0
71	227	13	221	1937	·94	·06	8·0	8·5
72	214	15	207	1716	·93	·07	7·5	8·0
73	199	16	191	1509	·92	·08	7·1	7·6
74	183	16	175	1318	·91	·09	6·7	7·2
75	167	16	159	1143	·90	·10	6·3	6·8
76	151	15	143	984	·90	·10	6·0	6·5
77	136	15	129.	841	·89	·11	5·7	6·2
78	121	14	114	712	·88	·11	5·4	5·9
79	107	13	100	598	·88	·12	5·1	5·6
80	94	12	88	498	·87	·13	4·8	5·3
81	82	11	77·	410	·86	·14	4·5	5·0
82	71	10	66·	333	·86	·14	4·2	4·7
83	61	9	56·	267	·85	·15	3·9	4·4
84	52	8	48·	211	·84	·16	3·5	4·0
85	44	8	40	163	·82	·18	3·2	3·7
86	36	7	33·	123	·81	·19	2·9	3·4
87	29	6	26·	90	·79	·20	2·6	3·1
88	23	5	20	64	·78	·22	2·3	2·8
89	18	4	16·	44	·77	·23	1·9	2·4
90	14	4	12	28	·71	·29	1·4	1·9
91	10	4	8	16	·60	·40	1·0	1·5
92	6	3	5	8	·50	·50	0·7	1·2
93	3	2	2	3	·33	·67	0·3	0·8
94	1	1	1·	1·	·00	1·00	0·0	0·5

Note.—The figures in the foregoing table, and in those which follow at pages 82 and 90, are not carried out to a sufficient number of decimal places to give accurate results. They are merely intended to illustrate the method of procedure.

(3) From the foregoing table find the probability of a person aged 71 living 10 years.

$$_{10}p_{71} = \frac{l_{81}}{l_{71}} = \frac{82}{227} = \cdot 361.$$

(4) Find the probability that a person aged 71 will die within 10 years.

$$/_{10}q_{71} = \frac{d_{71} + d_{72} + \ .\ .\ . + d_{80}}{l_{71}}$$

$$= \frac{l_{71} - l_{81}}{l_{71}} = \frac{227 - 82}{227} = \cdot 639.$$

This probability might also be found by deducting from 1 the probability of a person aged 71 living 10 years, thus

$$1 - {}_{10}p_{71} = 1 - \cdot 361 = \cdot 639,$$

which corresponds with the result already found.

(5) Find the probability of a person aged 61 and another aged 80, both living for 5 years.

$$_{5}p_{61:80} = \frac{l_{66:85}}{l_{61:80}} = \frac{290 \times 44}{350 \times 94} = \cdot 388.$$

(6) Find from the foregoing table the number of pensioners aged 75 and upwards in a society to which 50 persons of age 30 are yearly admitted. Given $l_{30} = 564.$

$$\frac{50}{l_{30}} \times T_{75} = \frac{50}{564} \times 1143 = 101.$$

CHAPTER IV.

COMMUTATION COLUMNS,

THEIR USE AND MODE OF CONSTRUCTION.

IN connection with the various mortality tables, sets of what are known as *Commutation Columns* have been prepared, by the use of which the calculation of life benefits of every description is greatly simplified.

Commutation Columns.

These columns have no meaning in themselves. They were contrived simply as aids to calculation, and received their name on account of the facility with which, by employing them, one benefit can be commuted for another.

There are six *Commutation Columns* in common use, known respectively as the D_x, N_x, S_x, C_x, M_x, and R_x columns, and these are tabulated for different rates of interest.

The D_x column gives for each age (down to the limit of life) the products of v^x and l_x.

The N_x column, as ordinarily constructed, gives opposite each age the sum of the values of D_x from one year older to the end of the table. Sometimes a slightly different form of N_x column is used, but this will more appropriately be considered in connection with Life Annuities.

The C_x column contains the products of v^{x+1} and d_x for each age, and the M_x column gives for each age the sum of the values of C_x from that age to the limit of life.

The S_x and R_x columns are obtained by summing the values of N_x and M_x respectively; but for the simple benefits we shall here treat of these columns are not required.

We have accordingly

$$D_x = v^x l_x \qquad\qquad C_x = v^{x+1} d_x$$

$$N_x = D_{x+1} + D_{x+2} + \text{etc.} \qquad M_x = C_x + C_{x+1} + \text{etc.}$$

$$S_x = N_x + N_{x+1} + \text{etc.} \qquad R_x = M_x + M_{x+1} + \text{etc.}$$

Construction
of N_x and D_x
Columns.
To construct N_x and D_x columns for a given mortality table we proceed as follows:—

The ages to the limit of life are placed in a vertical column; in a second the number living at each age, and in a third the power of v corresponding to each age calculated at the rate of interest at which the values are to be tabulated. The amounts in columns 2 and 3 are then multiplied together, and the products placed in column 4. These products form the values of D_x. Then beginning at the last age in the table, the numbers in column 4 are summed consecutively, the total from D_{x+1} onwards being placed opposite age x, and the values of the N_x column are thus obtained.

Construction
of C_x and M_x
Columns.
To construct C_x and M_x columns the method of proceeding is very similar.

The ages to the limit of life being arranged in a vertical column, in a second the numbers who die at each age are placed, and in a third the value of the corresponding powers of v, the index in each case being higher by 1 than the age. The numbers in columns 2 and 3 are then multiplied together, and the products, which are placed in column 4, give the values of C_x. The M_x column is then formed by beginning at the last age in the table, adding consecutively the values of C_x, and placing the sums opposite the respective ages.

A practical illustration of the formation of D_x, N_x, C_x, and M_x columns will be found at pages 82 and 90.

Annuity and
Assurance
Columns.
D_x, N_x, and S_x are sometimes called the annuity columns, and C_x, M_x, and R_x the assurance columns. There is a connection between these which it may be well to point out.

We have

$$C_x = v^{x+1}d_x = v^{x+1}l_x - v^{x+1}l_{x+1} = vD_x - D_{x+1},$$

and similarly

$$M_x = vN_{x-1} - N_x, \quad \text{and} \quad R_x = vS_{x-1} - S_x.$$

Commutation columns are also used for joint life benefits, although a slight alteration is necessarily made on their form.

CHAPTER V.

LIFE BENEFITS.

ENDOWMENTS AND LIFE ANNUITIES.

Value of
Endowment. THE simplest form of benefit dependent on the contingencies of life is an *Endowment*, by which is meant a sum payable to an ·individual at a future date, provided he survive.

If the payment were in any event due at the end of the period— say n years—its value would be v^n, but, as it only becomes payable if the recipient survive the term of payment, it is necessary, in order to give effect to this contingency, to multiply by the probability that he will live for n years.

The value of an endowment of 1 is accordingly

$$v^n{}_np_x, \text{ or } \frac{v^n l_{x+n}}{l_x}.$$

We may also look at the matter from a slightly different point of view.

Let us suppose that there are l_x persons of age x, and that to each of them who survives n years a payment of 1 has to be made, the aggregate amount of these payments will be l_{x+n}, and their present value $v^n l_{x+n}$. This being the value of the total sum payable, the share of each of the l_x persons will on the average be

$$\frac{v^n l_{x+n}}{l_x}.$$

This method of finding the value of an endowment emphasizes the fact that the calculated value of any life benefit, while thoroughly to be depended on if we are dealing with a sufficient number of persons to constitute an average, may be very wide of the mark in an individual and isolated case.

The symbol for an *Endowment* payable in n years is $_nE_x$.

Accordingly
$$_nE_x = v^n {}_np_x = \frac{v^n l_{x+n}}{l_x}.$$

Endowment in Commutation Symbols. From the above formulae the value of an endowment could be calculated by taking the required values of l_x or p_x from the mortality table. If, however, the *Commutation Columns* referred to in the last chapter are made use of the value is obtained much more easily, thus

$$_nE_x = \frac{v^n l_{x+n}}{l_x},$$

or, multiplying the numerator and denominator by v_x^x;

$$= \frac{v^{x+n} l_{x+n}}{v^x l_x} = \frac{D_{x+n}}{D_x}.$$

By taking, therefore, the number in the D_x column opposite age $x+n$ and dividing it by the number opposite age x, the value of the endowment is at once obtained.

Value of Life Annuity. We pass now from this simplest form of benefit, to consider *Life Annuities*, which may be said to form the groundwork of all calculations involving the contingencies of life.

If the different instalments of a life annuity of 1 were certainly payable, their present value—as in the case of an annuity-certain—would be represented by the series

$$v + v^2 + v^3 + \text{etc.},$$

but, as the respective payments only become due in the event of the annuitant being alive to receive them, each term must be multiplied by the value of this probability.

If, therefore, the probability of a person aged x living one year be represented by p_x, of his living two years by $_2p_x$, and so on, the value of a life annuity of 1 will be

$$vp_x + v^2 {}_2p_x + v^3 {}_3p_x + \text{etc.}$$

It will be noted that a life annuity is thus a series of endowments.

The probability that the annuitant will be alive when the payments fall due is manifestly a quantity which year by year diminishes in value, and at length, when the limiting age is reached, vanishes altogether. Whenever this factor becomes 0, the series comes to an end.

To find the value of the above series, which thus continues till the limit of life is reached, let us represent the value of a *Life Annuity* of 1 by the symbol a_x, and we have

$$a_x = vp_x + v^2{}_2p_x + v^3{}_3p_x + \text{etc.}$$

$$= v\frac{l_{x+1}}{l_x} + v^2\frac{l_{x+2}}{l_x} + v^3\frac{l_{x+3}}{l_x} + \text{etc.}$$

$$= \frac{v\,l_{x+1} + v^2\,l_{x+2} + v^3\,l_{x+3} + \text{etc.}}{l_x}.$$

Multiply the numerator and denominator of this fraction by v^x, which of course in no way changes its value, and

$$a_x = \frac{v^{x+1}l_{x+1} + v^{x+2}l_{x+2} + v^{x+3}l_{x+3} + \text{etc.}}{v^x l_x}.$$

It will be noted that the indices of v are now in every case the same as the suffixes of l. Accordingly for $v^x l_x$ the commutation symbol D_x may be used, and we have

$$a_x = \frac{D_{x+1} + D_{x+2} + D_{x+3} + \text{etc.}}{D_x}.$$

If for the series $D_{x+1} + D_{x+2} + \text{etc.}$ (the series being continued to the limiting age of the table) we now write N_x, we have

$$a_x = \frac{N_x}{D_x}.$$

Accordingly, when D_x and N_x columns have been constructed by the method described in the last chapter, to find the value of a life annuity at any age it is only necessary to divide the amount in the N_x column by the corresponding value of D_x.

Different Methods of forming the Nx column. It will be noted that, in order to make the expression for a Life Annuity symmetrical, N_x is formed by summing the D_x column from age $(x+1)$ onwards.

While this form of the N_x column is usually adopted, it is not universal. Some actuaries, such for example as Dr. Farr in the English Life Tables, construct it by summing, not from D_{x+1}, but from D_x. Care must therefore be taken before using annuity tables to ascertain in which of these ways the values of N_x have been obtained. This is easily done by looking at the end of the table. If the N_x column has been formed by summing from D_{x+1}, the value in it opposite the last age will be 0, but if it has been constructed by summing from D_x, the value at the last age will be the same as that in the D_x column.

If the tables be constructed in the latter form the value of an annuity is

$$\frac{N_{x+1}}{D_x}.$$

Life Annuity not of same value as Annuity-Certain for Expectation of Life. It is a common error to suppose that a life annuity is of the same value as an annuity-certain for the term of the curtate expectation of life. In each case the total amount payable will on the average be the same, but the present value of the life annuity is less than that of the annuity-certain, as the payments of the former (continuing as they do till the limit of life is reached) are manifestly spread over a longer period, and more affected by discount than those of the annuity-certain.

Annuities-due As in the case of annuities-certain, it is assumed (unless otherwise stated) that the first payment of a life annuity is due at the end of the first period. If the first payment of a life annuity be due at the beginning of the first year, the annuity is called an *Annuity-due.* For the value of an annuity-due we have

$$a_x = 1 + a_x,$$

or, in commutation symbols

$$a_x = \frac{N_{x-1}}{D_x}.$$

To find the value of an annuity-due, we thus divide the value in the N_x column at one year less than the given age by the value of D_x at the given age. If, however, the tables be in the form adopted by Dr. Farr and others, then

$$a_x = \frac{N_x}{D_x}.$$

We shall hereafter assume that the N_x column is constructed in the usual form.

In the foregoing formulae for the values of annuities
Complete
Annuity. no account has been taken of the $\frac{1}{2}$, which will, on the
average, be payable in the year of death if the annuity
be apportionable. When this is allowed for the annuity is called a
Complete Annuity. If the value of 1 payable at the end of the year
of death be represented by A_x, it is, for most practical purposes, a
sufficiently close approximation to take as the value of a *Complete
Annuity* of 1

$$a_x + \frac{A_x}{2}.$$

If, however, a more accurate result be desired, it may be found
from the formula

$$\overset{\circ}{a}_x = a_x + \frac{A_x}{2}(1 + i)^{\frac{1}{2}}.$$

Let us now consider briefly the value of a *Deferred Annuity.*

An annuity deferred n years is one which will be
Deferred
Annuity. entered upon in n years, and the first payment of
which will be made at the end of the $(n+1)^{th}$ year if
the annuitant be then alive.

At the end of n years the value of this annuity will be a_{x+n}, but
to find its present value, it is necessary to multiply by v^n, and also
by $_np_x$, *i.e.* the probability of a person now aged x being alive
at the end of n years to enter upon the annuity.

For the value of an annuity deferred n years the symbol is $_n|a_x$;
and, accordingly

$$_n|a_x = v^n\,_np_x \times a_{x+n}$$

To express this value in commutation symbols

$$v^n\,_np_x = \frac{v^{x+n}l_{x+n}}{v^x l_x} = \frac{D_{x+n}}{D_x}, \text{ and } a_{x+n} = \frac{N_{x+n}}{D_{x+n}}.$$

Therefore

$$_n|a_x = \frac{D_{x+n}}{D_x} \times \frac{N_{x+n}}{D_{x+n}} = \frac{N_{x+n}}{D_x}.$$

To find the value of an annuity of 1 deferred n years it is thus only
necessary to divide the value of N_{x+n} by D_x.

Temporary Annuities. A *Temporary Annuity* is equivalent to a whole life annuity, less a deferred annuity.

For the value of a temporary annuity for n years to a person aged x the symbol is $/_n a_x$, and accordingly

$$/_n a_x = a_x - {}_n/a_x,$$

or, in commutation symbols

$$/_n a_x = \frac{N_x}{D_x} - \frac{N_{x+n}}{D_x} = \frac{N_x - N_{x+n}}{D_x}.$$

To find the value of a temporary annuity of 1 for n years we accordingly subtract N_{x+n} from N_x and divide the remainder by D_x.

Joint Life Annuities. So far the formulae we have investigated have referred to one life only, but the demonstrations apply with equal force to *Joint Life Annuities.*

Accordingly

$$a_{xy} = \frac{v l_{x+1 : y+1} + v^2 l_{x+2 : y+2} + \text{etc.}}{l_{xy}}.$$

Reversionary Annuities. An annuity to commence on the death of a person now aged y, and to continue thereafter so long as another now aged x may live, is called a *Reversionary Annuity,* and is denoted by the symbol $a_{y/x}$.

This benefit is manifestly represented by the payments of an annuity on (x) which will fall due after (y) dies, and its value is therefore that of an annuity on (x) less the value of an annuity while both are alive.

Accordingly

$$a_{y/x} = a_x - a_{xy}.$$

Life Interest. A *Life Interest* is a benefit very similar to an annuity, the difference being that while the latter has a fixed annual value, the amount derived from the former may vary from year to year. In valuing a life interest it is necessary to average the annual income, and take it as a fixed quantity, when the benefit becomes practically a life annuity, and is treated accordingly.

Practical Examples.

(1) Given the l_x column as undernoted, construct D_x and N_x columns and also a Table of Life Annuities at 3 per cent. interest.

Age x	l_x	v^x@3%	D_x	N_x	a_x	Age x	l_x	v^x@3%	D_x	N_x	a_x
1	2	3	4	5	6	1	2	3	4	5	6
30	564	·412	232	4550	19·6	63	326	·155	50·5	483·5	9·6
31	559	·400	224	4326	19·3	64	314	·151	47·4	436·1	9·2
32	554	·388	215	4111	19·1	65	302	·146	44·1	392·0	8·9
33	549	·377	207	3904	18·9	66	290	·142	41·2	350·8	8·5
34	544	·366	199	3705	18·6	67	278	·138	38·4	312·4	8·1
35	538	·355	191	3514	18·4	68	266·	·134	35·6	276·8	7·8
36	532	·345	184	3330	18·1	69	253	·130	32·9	243·9	7·4
37	526	·335	176	3154	17·9	70	240	·126	30·2	213·7	7·1
38	520	·325	169	2985	17·7	71	227	·123	27·9	185·8	6·7
39	514	·316	162	2823	17·4	72	214	·119	25·5	160·3	6·3
40	508	·306	155	2668	17·2	73	199·	·115	22·9	137·4	6·0
41	502	·298	150	2518	16·8	74	183	·112	20·5	116·9	5·7
42	496	·289	143	2375	16·6	75	167·	·109	18·2	98·7	5·4
43	490	·280	137	2238	16·3	76	151	·106	16·0	82·7	5·2
44	483	·272	131	2107	16·1	77	136	·103	14·0	68·7	4·9
45	476	·264	126	1981	15·7	78	121	·100	12·1	56·6	4·7
46	469	·257	121	1860	15·4	79	107	·097	10·4	46·2	4·4
47	462	·249	115	1745	15·2	80	94	·094	8·8	37·4	4·2
48	455	·242	110	1635	14·9	81	82·	·091	7·5	29·9	4·0
49	448	·235	105	1530	14·6	82	71	·088	6·3	23·6	3·7
50	441	·228	100	1430	14·3	83	61	·086	5·2	18·4	3·5
51	434	·221	96	1334	13·9	84	52	·083	4·3	14·1	3·3
52	427	·215	92	1242	13·5	85	44	·081	3·6	10·5	2·9
53	420	·209	88	1154	13·1	86	36	·079	2·9	7·6	2·6
54	413	·203	84	1070	12·7	87	29	·076	2·2	5·4	2·4
55	406	·197	80	990	12·4	88	23·	·074	1·7	3·7	2·2
56	398	·191	76	914	12·0	89	18	·072	1·3	2·4	1·8
57	390	·185	72	842	11·7	90	14	·070	1·0	1·4	1·4
58	382	·180	69	773	11·2	91	10	·068	0·7	0·7	1·0
59	373	·175	65	708	10·9	92	6	·066	0·4	0·3	·8
60	362	·170	62	646	10·4	93	3	·064	0·2	0·1	·5
61	350	·165	58	588	10·1	94	1	·062	0·1		
62	338	·160	54	534	9·9	95	0	·060			

(2) On the basis of the foregoing table, what is the value of a sum of £1000 payable in 20 years to a person presently aged 30 provided he survive?

$$1000 \times {}_{20}E_{30} = 1000 \times \frac{D_{50}}{D_{30}}$$

$$= 1000 \times \frac{100}{232} = £431 \ 0s. \ 8d.$$

(3) What on the basis of this table is the value of an Annuity of £20 deferred 15 years to a person presently aged 35?

$$20 \times {}_{15/}a_{35} = 20 \times \frac{N_{50}}{D_{35}}$$

$$= 20 \times \frac{1430}{191} = £149 \ 14s. \ 9d.$$

(4) What on the same basis is the value of a Temporary Annuity of £20 for 20 years to a person aged 31?

$$20 \times {}_{/20}a_{31} = 20 \times \frac{N_{31} - N_{51}}{D_{31}}$$

$$= 20 \times \frac{4326 - 1334}{224} = £267 \ 2s. \ 10d.$$

(5) A property yielding an average annual return of £100, life-rented by a person aged 45, is sold for £2500 by mutual arrangement between the life-renter and the reversioner. How should the price realised be apportioned between them?

If the income be divided by the price realised the rate of interest which the capital value of the property has been yielding is found. In this case

$$\frac{100}{2500} = \cdot 04, \ or \ 4 \ per \ cent.$$

The share of the life-renter is then obtained by multiplying the income by the value of a complete annuity on his life at this rate of interest, *i.e.*

$$100 \times \overset{\circ}{a}_{45} \quad (\overset{\circ}{a}_{45} \ being \ taken \ at \ 4 \ per \ cent. \ interest).$$

The reversioner's share is the balance of the price.

This is known as Baden's method of apportionment, and in cases such as the above is now generally adopted.

CHAPTER VI.

LIFE BENEFITS.

ASSURANCES AT SINGLE AND ANNUAL PREMIUMS.
POLICY VALUES.

Connection between Annuities and Assurances. BETWEEN the values of annuities and assurances there is a close connection, and if the value of an annuity on the life of any individual be given, it is easy without further data to find the value of a sum payable on his death.

To illustrate this, let us suppose that the amount insured is 1. If the assurance were payable forthwith its value would be 1, but as it does not become due till the assured dies it is necessary, in order to determine its value, to deduct from 1 the value of an annuity of the interest upon it during his life. The value of an assurance of 1 on the life of a person aged x is, therefore, 1 minus the value of an annuity of the interest upon it for every year on which the assured enters, and as the annual interest on 1 payable in advance is iv or d, writing A_x for the value of the assurance, we have

$$A_x = 1 - d(1 + a_x). \qquad (1)$$

From this equation we derive the further relations

$$A_x = v(1 - ia_x) = \frac{1 - ia_x}{1 + i}, \qquad (2)$$

$$\text{and, } A_x = v(1 + a_x) - a_x. \qquad (3)$$

These results can also be obtained by direct reasoning.

By transposing the above equations the value of an annuity can be expressed in terms of the corresponding assurance, thus

$$a_x = \frac{1 - A_x}{d} - 1, \qquad (1)$$

$$a_x = \frac{v - A_x}{d}, \qquad (2)$$

and, $\quad a_x = \frac{1 - (1 + i)A_x}{i}. \qquad (3)$

Value of Assurance deduced from Mortality Table. Let us now consider how the value of a *Life Assurance* may be deduced from the mortality table by a direct process.

If the lives of l_x persons of age x be assured, each for 1, the amount to be paid at the end of the first year will be d_x; at the end of the second year d_{x+1}; at the end of the third year d_{x+2}, and so on, the total amount payable in respect of the l_x lives being represented by the series

$$d_x + d_{x+1} + d_{x+2} + \text{etc.}$$

The aggregate present value of these assurances is .

$$vd_x + v^2d_{x+1} + v^3d_{x+2} + \text{etc.}$$

and the average present value of an assurance of 1 on any one of the l_x persons

$$\frac{vd_x + v^2d_{x+1} + v^3d_{x+2} + \text{etc.}}{l_x}.$$

If now the numerator and denominator of this fraction be multiplied by v_x, we have as the value of the assurance

$$A_x = \frac{v^{x+1}d_x + v^{x+2}d_{x+1} + v^{x+3}d_{x+2} + \text{etc.}}{v^x l_x}.$$

To express this value in commutation symbols

$$C_x = v^{x+1}d_x, \; D_x = v^x l_x, \text{ and } M_x = C_x + C_{x+1} + C_{x+2} + \text{etc.}$$

Whence $\qquad A_x = \dfrac{C_x + C_{x+1} + C_{x+2} + \text{etc.}}{D_x}$

$$= \frac{M_x}{D_x}.$$

To find the value of an assurance, it is thus only necessary to divide the value in the M_x column opposite the given age by the corresponding value in the D_x column.

Assumption as to time when Assurance Payable.
In arriving at the above formula for the value of a life assurance it has been assumed that it is payable at the end of the year of death, and, as on the average the deaths may be taken as occurring at the middle of the year, this will represent an assurance payable six months after death.

If the amount be payable at any other interval the formula must be modified accordingly. Thus an assurance payable at the moment of death is represented approximately by

$$A_x(1+i)^{\frac{1}{2}}.$$

Deferred Assurance.
A *Deferred Assurance* is one in which the risk does not commence to run till the expiry of a stipulated period—say n years.

To ascertain the value of a *Deferred Assurance*, the symbol for which is $_n/A_x$, the same reasoning is employed as in the case of a deferred annuity (see page 80), and accordingly we have

$$_n/A_x = v^n{}_np_x \times A_{x+n},$$

or, expressed in commutation symbols

$$= \frac{D_{x+n}}{D_x} \times \frac{M_{x+n}}{D_{x+n}}$$

$$= \frac{M_{x+n}}{D_x}.$$

Temporary Assurance.
A *Temporary* or *Term Assurance*, the symbol for which is $/_nA_x$, is equivalent to a whole life assurance less a deferred assurance.

Accordingly

$$/_nA_x = A_x - {}_n/A_x,$$

or, in commutation symbols

$$= \frac{M_x}{D_x} - \frac{M_{x+n}}{D_x}$$

$$= \frac{M_x - M_{x+n}}{D_x}.$$

These expressions are all analogous to those already found for the values of annuities.

Absolute Reversion. An *Absolute Reversion*, by which is meant a reversionary interest free from all conditions as to survivorship, is practically equivalent to an assurance payable on the death of the life-renter, and may be valued accordingly. The present value of each unit will thus be A_x, where x represents the age of the life tenant.

Single and Annual Premiums. A_x—the present value of an assurance of 1 on the life of a person aged x—is manifestly the *Net Single Premium* required for such assurance. The corresponding *Annual Premium* (payable in advance) may be found as follows :—

The value of an annuity-due of 1 being $(1 + a_x)$, 1 is the value of an annuity-due of $\dfrac{1}{1 + a_x}$. Accordingly the annuity-due which is equivalent to a single payment of A_x is

$$\frac{A_x}{1 + a_x}.$$

For the *Net Annual Premium* for an assurance of 1 we therefore have

$$P_x = \frac{A_x}{1 + a_x}.$$

To express the annual premium in commutation symbols

$$A_x = \frac{M_x}{D_x}, \text{ and } (1 + a_x) = \frac{N_{x-1}}{D_x}.$$

Whence

$$P_x = \frac{A_x}{1 + a_x} = \frac{M_x}{D_x} \times \frac{D_x}{N_{x-1}} = \frac{M_x}{N_{x-1}}.$$

If, therefore, a set of commutation columns be available, the *Net Single Premium* for an assurance of 1 on a person aged x is found by dividing M_x by D_x, and the corresponding *Annual Premium* by dividing M_x by N_{x-1}.

If the premium, instead of being paid throughout life, be only payable for a limited number of years, the value of a temporary annuity-due must be inserted in the formula, thus

$$_nP_x = \frac{M_x}{D_x} \times \frac{D_x}{N_{x-1} - N_{x+n-1}}$$

$$= \frac{M_x}{N_{x-1} - N_{x+n-1}}.$$

Similarily, if the assurance as well as the premium be temporary, the annual premium will obviously be

$$\frac{M_x - M_{x+n}}{N_{x-1} - N_{x+n-1}}$$

The premiums we have been considering are known as *Net Premiums*, to distinguish them from the *Office Premiums* charged by insurance companies, to which a *loading* (as it is technically called) has been added for expenses.

In making their valuations, insurance offices discard the loading, and value on the basis of the net premiums, and we shall now briefly consider what on this footing is meant by the *Reserve Value* of a Policy.

Reserve Values.

When the insurance is first effected the value of the net premiums to be received by the company should be exactly equal to that of the assurance granted. Now the present value of the net premiums for an assurance of 1 on a person aged x is $P_x(1 + a_x)$, and the value of the assurance is A_x, so that at the outset we have the equation

$$P_x(1 + a_x) = A_x.$$

It will be noted that this equation might have been derived algebraically from the above expression for the value of the annual premium.

At this stage the policy has no *reserve value*, or in other words the company's liability under its contract is exactly balanced by the value of the corresponding asset, provided a sufficient number are insured to give average results.

Let us now suppose that n years have elapsed since the policy was taken out, and consider how the different quantities are affected.

A_x—the value of the assurance—has become A_{x+n}, and accordingly has increased in value, as the period at which it will become payable is n years nearer than it was originally. On the other hand the value of the future premiums to be received by the company has decreased from $P_x(1 + a_x)$ to $P_x(1 + a_{x+n})$.

The liability of the company under the policy is thus no longer covered by the value of the corresponding asset, and, in place of the equation with which we started, we now have

$$P_x(1 + a_{x+n}) + Reserve\ Value = A_{x+n},$$

whence,

$$Reserve\ Value = A_{x+n} - P_x(1 + a_{x+n}),$$

or, expressed in commutation symbols,

$$= \frac{M_{x+n} - P_x \times N_{x+n-1}}{D_{x+n}}.$$

This mode of calculating the reserve is known as the *prospective method* of valuation, because the result is found by deducting from the present value of the sum assured that of the premiums to be received.

Prospective and Retrospective Methods of Valuation.

There is another system of valuation known as the *retrospective method*, which, however, is not so frequently employed. The results obtained by it are precisely the same as by the prospective method, but they are arrived at by taking account of the past. The *Reserve Value* from this point of view is the accumulated value of the premiums received less that of the risk already incurred.

Surrender Value.

From what has been said it is evident that the *Reserve Value*—$\{A_{x+n} - P_x(1 + a_{x+n})\}$—is the average amount by which an insurance office will profit by an assurance being cancelled, and accordingly it forms in theory the *Surrender Value* of the policy. In practice certain modifications require to be made on this value, but these we need not here consider.

Paid-up Policy.

In lieu of a cash surrender value a *paid-up policy* is often granted.

As the value of an assurance of 1 at age $x+n$ is A_{x+n}, the paid-up policy that can be given at that age in place of a cash surrender value, say of V, is

$$\frac{V}{A_{x+n}}.$$

Value of Policy to a Purchaser. It is hardly necessary to point out that the reserve value of a policy as investigated above does not represent its value to a purchaser. While an insurance company in valuing its policies deals only with the *net premiums* it receives, a purchaser has to pay the full *office premiums*. If, therefore, the above formula is to be made available for determining the value of a policy to a purchaser, the actual premium payable must be substituted for P_x.

Practical Examples.

(1) From the l_x column, as undernoted, construct C_x and M_x columns at 3 per cent. interest, and also tables of Net Single and Annual Premiums for an Assurance of 1.

Age x	l_x	d_x	v^{x+1} @ 3°/$_\circ$	C_x	M_x	D_x	A_x	N_{x-1}	P_x
1	2	3	4	5	6	7	8	9	10
30	564	5	·400	2·000	92·369	232	·398	4782	·0193
31	559	5	·388	1·940	90·369	224	·403	4550	·0199
32	554	5	·377	1·885	88·429	215	·411	4326	·0204
33	549	5	·366	1·830	86·544	207	·418	4111	·0211
34	544	6	·355	2·130	84·714	199	·426	3904	·0217
35	538	6	·345	2·070	82·584	191	·432	3705	·0223
36	532	6	·335	2·010	80·514	184	·438	3514	·0229
37	526	6	·325	1·950	78·504	176	·446	3330	·0236
38	520	6	·316	1·896	76·554	169	·453	3154	·0243
39	514	6	·306	1·836	74·658	162	·461	2985	·0250
40	508	6	·298	1·788	72·822	155	·470	2823	·0258
41	502	6	·289	1·734	71·034	150	·474	2668	·0266
42	496	6	·280	1·680	69·300	143	·485	2518	·0275
43	490	7	·272	1·904	67·620	137	·494	2375	·0285
44	483	7	·264	1·848	65·716	131	·502	2238	·0294
45	476	7	·257	1·799	63·868	126	·507	2107	·0303
46	469	7	·249	1·743	62·069	121	·513	1981	·0313
47	462	7	·242	1·694	60·326	115	·525	1860	·0324
48	455	7	·235	1·645	58·632	110	·533	1745	·0336
49	448	7	·228	1·596	56·987	105	·543	1635	·0349

Note.—The value of M_{49} is obtained by summing the values of C_x from the limiting age to age 49 inclusive.

(2) From the above table find the value of a Temporary Assurance of £1000 for 10 years on the life of a person aged 36.

$$1000 \times {}_{/10}A_{36} = 1000 \times \frac{M_{36} - M_{46}}{D_{36}}$$

$$= 1000 \times \frac{80 \cdot 514 - 62 \cdot 069}{184}$$

$$= £100 \text{ 4s. 11d.}$$

(3) According to this table what is the Net Annual Premium for the above Assurance ?

The premium required is

$$1000 \times \frac{M_{36} - M_{46}}{N_{35} - N_{45}} = 1000 \times \frac{80 \cdot 514 - 62 \cdot 069}{3514 - 1981}$$

$$= 1000 \times \frac{18 \cdot 445}{1533}$$

$$= £12 \text{ 0s. 8d.}$$

(4) A person aged 35 insures his life for £1000. What amount would the same Single Premium have secured if the Assurance had been deferred 10 years ?

∘Writing x for the required amount, we have

$$1000 \times \frac{M_{35}}{D_{35}} = x \times \frac{M_{45}}{D_{35}}$$

$$\text{whence,} \quad x = 1000 \frac{M_{35}}{M_{45}}$$

$$= 1000 \times \frac{82 \cdot 584}{63 \cdot 868}$$

$$= £1293 \text{ 0s. 10d.}$$

(5) Given $a_{20} = 20 \cdot 245$ at $3\frac{1}{2}$ per cent. interest. Find A_{20}.

To deduce the value of an assurance from that of the corresponding annuity three formulae were investigated. Let us take these in turn—

(a)
$$A_{20} = 1 - d(1 + a_{20})$$
$$= 1 - \cdot 0338 \times 21 \cdot 245 \qquad = \cdot 282.$$

(b)
$$A_{20} = v(1 - ia_{20})$$
$$= \cdot 9662(1 - \cdot 035 \times 20 \cdot 245) = \cdot 282.$$

(c)
$$A_{20} = v(1 + a_{20}) - a_{20}$$
$$= \cdot 9662 \times 21 \cdot 245 - 20 \cdot 245 = \cdot 282.$$

(5) A Life Assurance Company has granted policies of insurance to various persons as follows :—

$$£20,000 \text{ taken out at age } 30,$$
$$5,000 \quad ,, \qquad ,, \quad 33,$$
$$10,000 \quad ,, \qquad ,, \quad 37,$$
$$12,000 \quad ,, \qquad ,, \quad 42.$$

The present age of all these persons is 48. Valuing on the basis of the foregoing table, and assuming that the above ages are exact, and that accordingly no adjustments require to be made for fractional parts of a year, what reserve does the Company require to hold in respect of these policies?

The *Net Premiums* annually received will be

$$20,000 \times P_{30} = 20,000 \times \cdot0193 = 386\cdot0$$
$$5,000 \times P_{33} = 5,000 \times \cdot0211 = 105\cdot5$$
$$10,000 \times P_{37} = 10,000 \times \cdot0236 = 236\cdot0$$
$$12,000 \times P_{42} = 12,000 \times \cdot0275 = 330\cdot0$$

$$£1057\cdot5$$

If the premiums had all been paid immediately before the date of the valuation, the value of the future premiums would have been represented by an annuity, but if, on the other hand, they were payable immediately thereafter, their value would constitute an annuity-due. As some of the premiums would be payable before and others after the valuation, they may on the average be taken at the mean between an annuity and an annuity-due, *i.e.* $(\frac{1}{2} + a_x)$. The value of the net future premiums will accordingly be

$$1057\cdot5(\tfrac{1}{2} + a_{48}) = 1057\cdot5 \times (\cdot5 + 14\cdot9) = 16285\cdot5$$

and the value of the sums assured

$$47,000 \times A_{48} = \qquad 47,000 \times \cdot533 = 25051\cdot0$$

The *Reserve* required is therefore   £8765·5

I.

PRINCIPLES OF LOGARITHMIC CALCULATION.

IT is an elementary rule in algebra that to multiply powers of the same quantity we add the indices. Thus to multiply a^3 by a^2 the exponents, 3 and 2, are added, and a^5, the required product, is obtained. Similarly, to divide a^3 by a^2 the latter exponent is subtracted from the former, and we have as the result $a^{(3-2)}$, or a.

Further, if it be desired to raise a^2 to the third power, we multiply the index by 3, and so obtain a^6. And similarly, if it were required to extract the cube root of a^2, by dividing the index by 3, the required root $a^{\frac{2}{3}}$ is found at once.

Now it is easier to add than to multiply and to subtract than to divide, and much easier to multiply than to raise to a power and to divide than to extract a root. A system of calculation, therefore, by which multiplication can be performed by addition, and division by subtraction, involution by multiplication and evolution by division, will manifestly afford great economy of labour. This is the principle underlying calculation by logarithms, and the simple elementary rule to which we have referred is the key to the whole system.

To illustrate this let us take the following powers of 10 corresponding to the first ten natural numbers—

$1 = 10^{\cdot 000}$		$6 = 10^{\cdot 778}$	
$2 = 10^{\cdot 301}$		$7 = 10^{\cdot 845}$	
$3 = 10^{\cdot 477}$		$8 = 10^{\cdot 903}$	
$4 = 10^{\cdot 602}$		$9 = 10^{\cdot 954}$	
$5 = 10^{\cdot 699}$		$10 = 10^{1 \cdot 000}$	

With the help of these powers of 10 we can make calculations in the manner above indicated.

To multiply we add the powers—

$$2 \times 5 = 10^{.301 + .699} = 10^{1.000} = 10.$$

To divide we take the difference between them—

$$6 \div 2 = 10^{.778 - .301} = 10^{.477} = 3.$$

To raise to a power we multiply—

$$3^2 \quad = 10^{.477 \times 2} \quad = 10^{.954} = 9.$$

To extract a root we divide—

$$\sqrt[3]{8} \quad = 10^{.903 \div 3} \quad = 10^{.301} = 2.$$

This is the whole principle of calculation by logarithms, and in accordance with the above table ·301 is the *logarithm* of 2 to the *base* 10; ·477 the logarithm of 3 to the same base; ·602 the logarithm of 4, and so on. Logarithms to the base 10 are known as common logarithms.

The advantages of this method of calculation were perceived in very early times, but it was not till the beginning of the seventeenth century that the difficulties in the way of its application were overcome.

The merit of being the first to construct a practicable table of logarithms admittedly belongs to John Napier, Baron of Merchistoun, who, in 1614, printed his "Canon Mirabilis Logarithmorum." His tables, which were very incomplete, were founded on what is now known as the natural base—2·718281828—which is usually designated by the symbol *e*.

In actuarial calculations it is sometimes necessary to use logarithms to this natural base. It is easy, however, by means of a divisor, called the *modulus*, to deduce from the common logarithm of any number the corresponding logarithm to the Napierian base.

The *modulus* for the common system of logarithms is $\dfrac{1}{\log_e 10}$ or ·43429448. If the common logarithm of any number be divided by this *modulus* it is at once converted into the corresponding Napierian logarithm.

The base of the Napierian system, the modulus above mentioned, and the Napierian logarithms of 2, 3, and 5, have all been calculated to upwards of 200 places of decimals.

From what has been said it is manifest that from the logarithms of two or three numbers, the logarithms of many others can be deduced. Thus, for example, if the logarithms of $2 = \cdot301$, $7 = \cdot845$, and $9 = \cdot954$ be given, we can find the logarithms of all the other numbers in the above table as follows—

$\log 1 = 0.$ (any number raised to the power 0 being 1.)

$\log 2 = \cdot301$ (given).

$\log 3 = \log \sqrt{9} = \frac{1}{2} \log 9 = \cdot477.$

$\log 4 = \log 2^2 = 2 \log 2 = \cdot602.$

$\log 5 = \log (10 \div 2) = \log 10 - \log 2 = 1 - \cdot301 = \cdot699.$

$\log 6 = \log (2 \times 3) = \log 2 + \log 3 = \cdot778.$

$\log 7 = \cdot845$ (given).

$\log 8 = \log 2^3 = 3 \log 2 = \cdot903.$

$\log 9 = \cdot954$ (given).

$\log 10 = 1.$

We can also find the logarithms of such numbers as the following—

$\log 98 = \log (7^2 \times 2) = 2 \log 7 + \log 2 = 1\cdot991.$

$\log 65610 = \log (9^4 \times 10) = 4 \log 9 + \log 10 = 4\cdot816.$

For simplicity of illustration we have been working with three decimal places, but in practice logarithms must be taken out to six decimal places at least.

Full directions for using Tables of Logarithms are prefixed to the various tables. With a little practice they can be freely used, and the risk of error is much less than that involved in making lengthy calculations by the ordinary processes of arithmetic.

APPENDIX.

II.

TABLES FOR THE EXPRESSION OF SHILLINGS AND PENCE IN DECIMAL FORM.

I.—Decimals of £1 corresponding to a given number of Shillings.

1s.	= ·05	11s. =	·55
2s.	= ·10	12s. =	·60
3s.	= ·15	13s. =	·65
4s.	= ·20	14s. =	·70
5s.	= ·25	15s. =	·75
6s.	= ·30	16s. =	·80
7s.	= ·35	17s. =	·85
8s.	= ·40	18s. =	·90
9s.	= ·45	19s. =	·95
10s.	= ·50	20s. =	1·00

II.—Decimals of £1 corresponding to a given number of Pence.

1d.	= ·00416̇	7d. =	·02916̇
2d.	= ·00833̇	8d. =	·03333̇
3d.	= ·01250	9d. =	·03750
4d.	= ·01666̇	10d. =	·04166̇
5d.	= ·02083̇	11d. =	·04583̇
6d.	= ·02500	12d. =	·05000

www.ingramcontent.com/pod-product-compliance
Lightning Source LLC
Chambersburg PA
CBHW022145020726
47496CB00008B/2559